Airplane Rigging

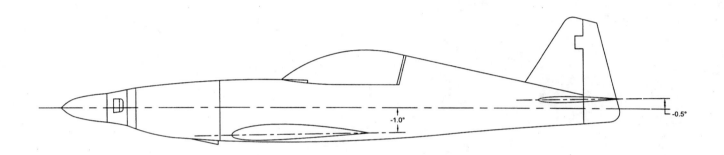

Authored, Compiled, and Edited
by Dave Russo
582 Rt 4A
Enfield, NH 03748

Published by:
Aircraft Technical Book Company
PO Box 270
Tabernash, CO 80478
970 887-2207
www.ACtechbooks.com

Airplane Rigging; 1st edition

ISBN 0-9774896-5-5

© 2007   Aircraft Technical Book Company
All rights reserved

Except as permitted under the United States Copyright Act, no part of this publication may be reproduced or distributed in any form, or by any means, or stored in a database or retrieval system without prior permission from the publisher.

Send all inquiries to:
Aircraft Technical Book Company
PO Box 270
Tabernash CO 80478
970 887-2207

www.ACtechbooks.com   (for aircraft maintenance technicians)
www.PilotsBooks.com   (for pilots and flight instructors)
www.BuildersBooks.com   (for custom aircraft builders)

Introduction

Aircraft rigging is a diverse subject that starts during construction of the aircraft. In it's most basic form, rigging makes an aircraft safe to fly. In terms of aircraft performance, rigging allows one to extract the most from the aircraft by reducing drag or changing performance in some specific flight regime. Like aircraft design, the rig is a compromise between its intended performance and controllability.

Small rigging changes can have drastic effects on performance, increasing it in some flight regimes while reducing performance in others, or at the very least make it easy or difficult to fly.

This manual is intended to continue where most leave off, and doesn't provide the basic information given in textbooks on aircraft construction and maintenance. It does provide some answers to the many questions that arise when building or rigging an airplane, and at the very least provides a path to the scientific method of determining an answer.

For the Homebuilder

The techniques, tools, and procedures given in this book may be used both during and after the construction process. The information contained herein is intended to accompany the kit or plans for an aircraft and answers many of the questions that arise during work.

Experimenting with the rigging of an airplane may allow significant increases in performance and controllability in a particular phase of flight. Some of the tools and techniques given here produce more precision than might be needed in many situations, but are useful for taking repeatable measurements when experimenting with the rig.

For the Mechanic

Although much of this manual is devoted to experimenting with homebuilts, the maintenance technician of certificated aircraft may find the tools and techniques provided here to be useful in daily shop operations. Troubleshooting rigging problems can be a frustrating and drawn out process. A rigging change needs to be followed by a flight test to determine the results. Usually this requires the aircraft owner to participate and the process can drag on over days or weeks. The methods given here will result in the fastest solution (or most efficient compromise) to rigging problems if the process is fully understood, however, it may require some extra work initially checking and resetting the rigging.

Some Generalizations

Airplane rigging can be complicated because of the interaction of the various components. Changing one aspect of aircraft geometry will result in aerodynamic changes not directly related to the initial geometry change. By understanding the relationships involved, rigging can be accomplished in less time with less flight testing.

On a lot of aircraft, proper rigging is going to mean getting more performance. It's like free horsepower. Errors in rigging are analogous to the great weight spiral in airplane design. Small changes require other changes, setting off a chain reaction that requires much compensation. A little twist here requires a little roll to compensate, and a little yaw to compensate for that, and so on. Some of the techniques in this manual suggest more precision than might normally be used, but taken cumulatively the performance benefits may be worth it. Slight changes in rigging are the main reason that two identical rental aircraft have a ten knot difference in speed or a 150 FPM difference in climb rate. Not only may performance be improved, but control authority is increased. Even slight out of rig conditions can make instrument flight very tiring in small airplanes. Old airplanes may find some new performance with a careful re-rig.

Airplane rigging must be approached with caution. It is possible to significantly change the flight characteristics in an adverse way, particularly if changes are made without knowledge of:

- the rigging specifications,
- the previous rigging condition,
- the effect of changes on flight characteristics.

Many homebuilders and those making modifications or designing from scratch don't have detailed rigging information and much experimentation may be required to arrive at the desired result.

An understanding of certain relationships enhances safety and makes work easier. It is frequently taken for granted that an aircraft behaves a certain way, like the horizontal or vertical stabilizer won't stall, sufficient control authority will be available for different loadings (CG's), and there is enough aileron and rudder authority for taking off and landing in crosswinds (from any direction), and in turbulence. That all can be changed by rigging improperly.

Table of Contents

Chapter 1 Airplane Geometry and Definitions 1
   Flight Control Systems 1
   Aircraft Axes 1
   Airframe Drawings 2
      Center Line 3
      Reference Datum 3
      Stations 4
      Buttock Lines 4
      Water Lines 4
   Airplane Components and Definitions 4
      Powerplant 4
      Fuselage 4
      Airframe 5
      Wing 5
      Wing Panel 5
      Wing Section 5
      Wing Planform 5
      Wing Root and Wing Tip 5
      Aileron 6
      Vertical Stabilizer and Rudder 6
      Horizontal Stabilizer and Elevator 6
      Stabilator 6
      Trim Tab 6
      Servo and Anti-Servo tabs 7
      Spoilers 8
      Empennage 8
      Control Surface 8
      Biplane Parts 8
   Airfoil Geometry 9
      Chord Line 9
      Mean Camber Line 11
      Symmetrical Airfoil 11
      Asymmetrical Airfoil 11
      Zero Lift Line 11
   Incidence Angle 11
   Wash-out 12
   Thrust Line 13
   Dihedral 13
   Control Deflection 14
   Decalage (Biplanes) 16
   Stagger (Biplanes) 16
   Gap (Biplanes) 17
   Mean Aerodynamic Chord 17

Chapter 2 Aerodynamics and Flight Mechanics Affecting Rigging 21
   Vectors 21
   Aircraft Movement 23
      Rotation 24
      Translation 24

Straight and Level Flight	26
Accelerated and Unaccelerated Flight	26
Aircraft Coordinate Systems	26
Body-Axis Coordinate System	26
Wind-Axis Coordinate System	27
Earth-Axis Coordinate System	28
Moment	29
Aircraft Control	29
Control Force	30
Control Effectiveness	30
Control Response	30
Control Power	30
Control Authority	30
Control Sensitivity	30
Trim	31
Trim and Stability	32
Conventional Aircraft in Pitch Equilibrium (Longitudinal Stability)	32
Slip, Skid, and Sideslip	33
Inclinometer or Slip/Skid Indicator	34
Coefficients	34
Airfoils/Wing Sections	35
Symmetrical Airfoil	36
Asymmetrical Airfoil	36
Lift Curve Slope	37
Total Aerodynamic Force	37
Lift	39
Spanwise Lift Distribution	40
Lift Curve Slope of a Whole Wing	40
Drag	42
Washout	43
Effect of Aircraft Rotation on Lift and Drag	43
Stall	44
Deflection of Flaps on Wings	45
Effect of Center of Gravity on Aircraft Performance	47
Performance Aspects of CG	48
Controllability and Stability Aspects of CG	48
Effect of Propeller on Airplane Dynamics	50
Yawing Moments and Sideslip	50
Pitching Moments	53
Rolling Moments	55
Rigging for Propeller Phenomenon	55
Yaw	56
Pitch	57
Roll	57
Incidence Angle of the Wing	58
Incidence Angle of the Horizontal Stabilizer	61
Effect of Incidence Changes on Maneuvering	63
Minimum Flying Qualities to Be Investigated After Incidence Changes of the Wing or Tail	63
Stall of the Horizontal Stabilizer	64
Bungee and Spring Centering Systems for the Elevator Control Circuit	65

- Biplane Aerodynamics 65
- Deep Stall Phenomenon 65
- Chapter 3 Rigging Tools 67
  - Metrology for Rigging 67
    - Accuracy, Resolution, and Repeatability 67
    - Symmetric Distribution of Error 68
    - Length of Straightedges 68
  - Measuring Instruments 69
    - Levels 69
    - Protractors/Clinometers; Propeller Protractor, Protractor Head, or Angle Finder ... 73
    - Trammel Bar 75
    - Plumb Bob 75
  - Adapters and Aircraft Specific Rigging Tools 75
    - Dihedral Board 76
    - Incidence Board 76
    - Throw Board 78
    - Neutral Board 78
    - Control Locks 81
    - Thrustline Measurement Tools 81
  - Cable Tensiometer 83
  - Wire Tensiometer 83
  - Streamlined Flying Wire Tool 84
  - Torque Seal 84
  - Turnbuckle Holder 85
  - Surveying Transit and Similar Equipment for Rigging 85
  - Slip/Skid Indicator 85
    - Yaw String 86
  - Attitude Indicator 87
  - Turn Coordinator/Turn and Bank Indicator/Turn and Slip Indicator 87
  - Compass and Heading Indicator 88
- Chapter 4 Factors in Rigging 89
  - Control Friction 89
  - Control Cable Tension 90
    - Adjusting Cable Tensions 92
  - Control Wear or Sloppiness 92
  - Control Stops 93
  - Control System Operational Tests Per FAR 23.683 94
  - Gap Seals 94
  - General Control Checks 95
  - Hardware Encountered in Rigging 96
    - Rod Ends 96
    - Turnbuckles 97
    - Clevis Forks 98
    - Flying Wires/Tie Rods 99
    - Check or Jam Nuts 101
  - Stall Strips 101
  - Fixed Trim Tabs 102
  - Spades 104
  - Vortex Generators 104
  - Type Certificate Data Sheets 105

Setting Up	105
Chapter 5 Initial Rigging	**107**
Setting the Fixed Surfaces	107
Leveling the Fuselage	107
Setting the Vertical Stabilizer	109
Setting the Wing	111
Setting the Thrustline	115
Setting the Horizontal Stabilizer	118
Setting the Slip/Skid Indicator	118
Setting the Movable Surfaces	118
Aileron and Elevator	118
Elevator Trim	123
Rudder	123
Flaps	125
Setting the Other Trims	125
Bungee and Spring Centering Systems for Primary Flight Controls	125
Aileron/Rudder Interconnect	125
Wheel Fairings	125
Wing Tips	126
Setting Wire Tensions	126
Landing and Taxi Lights	128
Compass	128
Testing the Rig	129
Chapter 6 Correcting Rigging Problems	**131**
Straight and Level Flight	131
Bendable Trim Tabs	132
Stalls	133
Maneuvering Flight	133
High Speed Flight	134
Chapter 7 Vibration	**135**
Elements of Vibration	135
Amplitude	135
Period and Cycle	135
Frequency	136
Classes of Vibration	136
Natural Frequency	137
Damping and Loss Factor	141
Resonance	141
Vibration of Rotating Parts	142
Measurement of the Amplitude of Vibration	144
Displacement	144
Velocity	144
Acceleration	145
Relationship Between Displacement, Velocity, and Acceleration	145
Relationship between Amplitude Measurement and Frequency	145
Structural Fatigue	146
Vibration Dampeners/Isolators	146
Noise	146
Aircraft Vibration	146
Sources of Aerodynamic Vibration	147

  Aircraft Skin 148
  Cables, Hoses, Tubing, and Antennas 148
  Instrument Panels 148
  Engine Mounts and Suspension Systems 149
 Powerplant Vibration 150
  Engine Vibration 150
  Propeller Vibration 151
 Airframe/Powerplant Resonance 152
 Troubleshooting Powerplant Vibrations 152
Chapter 8 Powerplant Rigging 155
 Propellers 155
  Setting Blade Angle 155
  Static Balancing 156
  Blade Tracking 158
  Dynamic Balancing 160
 Engine 162
  Idle Speed 162
  Mixture 162
  Controls 162
  Turbo/Supercharger 163
 Propeller and Propeller Governor Settings 163
  Adjusting Ground Adjustable Propellers for Performance 164
Chapter 9 Aeroelasticity and Control Surface Mass Balancing 167
 Mass Balancing 170
  Static Mass Balancing 171
  Dynamic Mass Balancing 171
 Balancing Terminology and Mathematics 171
 Measuring the Balance and Attaching Weights 174
Chapter 10 Landing Gear Rigging and Vibration 179
 Wheel Alignment 179
  Toe 179
  Camber 180
  Caster and Trail 181
 Rigging the Gear 185
 Landing Gear Problems 190
  Wheels and Tires 190
  Brakes 191
  Landing Gear 191
  Other Problems 194
  Preloading Wheel Bearings (Tapered Roller Bearings) 194
Chapter 11 Biplane Rigging 195
 Level the Fuselage 195
 Wing Assembly 195
 Wing Rigging 197
  Rigging the Center Section 197
  Rigging the Wings 200
Appendix A Math for Rigging 205
 Finding the Chord of An Arc 205
 Trigonometric Functions 206
References 207

Chapter 1
Airplane Geometry and Definitions

This chapter discusses the geometric relationships of the various parts of an airplane. Their effect on flight characteristics is discussed in Chapter 2 and in following chapters. An overview of basic control systems and concepts is provided to familiarize the reader with the terminology.

Flight Control Systems

Flight controls are classified as either primary or secondary. Primary flight controls are those that are used to direct the airplane in pitch, roll, and yaw. Secondary flight controls are trim, high-lift devices (flaps), speed brakes, etc..

This manual concentrates on small airplanes with *reversible* flight controls. Reversible flight controls mean that the control stick/yoke/pedals are connected directly to the surface that they actuate through cables, pushrods, bellcranks, pulleys, etc.. Moving the surface will cause movement of the control to which it is attached, hence the term reversible. *Irreversible* flight controls are generally found on large airplanes that have hydraulically actuated controls (attempting to move the control surface from the outside has little or no effect on its' cockpit control). Much of the terminology and ideas given in this chapter give the appearance of being the same as that used on large airplanes, however, the application of these ideas can be quite different on large airplanes and is not discussed in this book.

Flight control systems may be further broken down into open-loop or closed-loop systems. Most reversible flight control systems are closed-loop systems meaning that the actuator for a particular control surface cannot be moved in any direction if the control surface to which it is attached is held in a fixed position. Many rudder-pedal systems on small airplanes are open-loop, being attached to springs rather than to each other.

Aircraft Axes

Figure 1-1 illustrates the aircraft axes for the purpose of describing the layout. For layout and design, the axes are selected to be whatever point is convenient to reference drawings to, for example a structural member that is straight and approximately aligned with the axis in question. Rigging specifications are then made in reference to these axes. Because they lie at least in part inside the aircraft, designers will specify more convenient places from which to take measurements, or even add a structural member(s) to aid in measurements. The flight axes are discussed in more detail in Chapter 2.

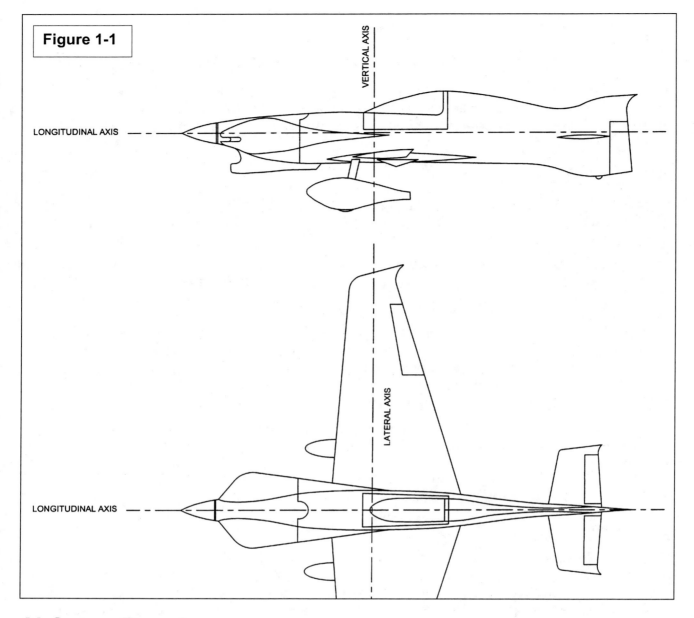

Figure 1-1

Airframe Drawings

Any point in or on an airframe can be identified by three coordinates. The coordinates are generally given in inches in relation to a fixed reference point. Aircraft drawings use particular terminology to identify these points. One may encounter some or all of the following terms in looking at the drawings or specifications. The exact wording and abbreviations have not always been strictly standardized, so it is likely one will encounter some variation of what is discussed here. Refer to Figure 1-2 for the following discussion.

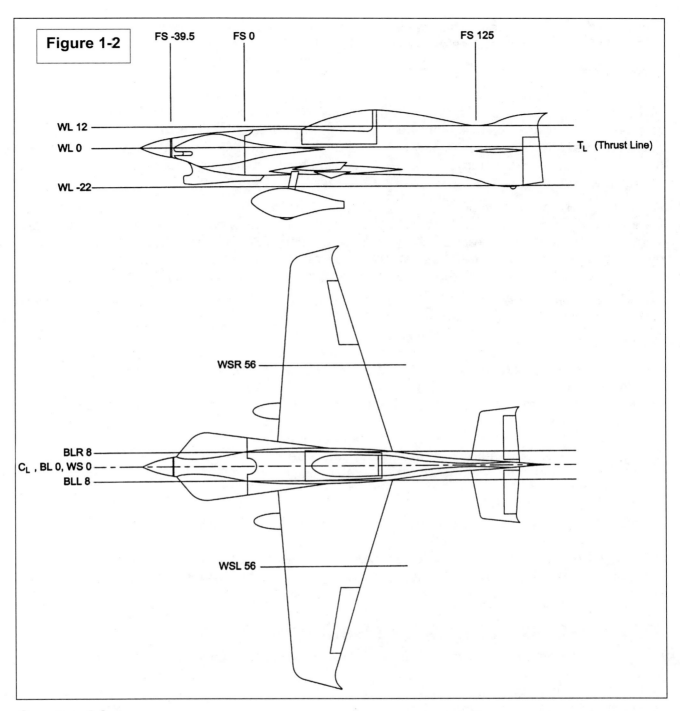

Figure 1-2

Center Line

The centerline, C_L, divides the aircraft into two symmetrical halves. It is the same as Buttock Line 0 (see Buttock Lines in the section), and is the longitudinal axis of the aircraft from which certain rigging measurements are made.

Reference Datum

The fixed reference point known as the reference datum is used to mark the zero location for measurements along the length of the fuselage. The aircraft designer will typically specify the location of the reference datum that is to be used for taking measurements. Often it is the forward side of the firewall. This may or may not be the same as the reference datum used for calculating weight and balance of the completed aircraft.

Stations

A fuselage station (FS or Sta.) is any point that is forward of, or behind the reference datum, along the longitudinal axis, and is usually measured in inches. Station numbers aft of the reference datum are positive numbers, and station numbers forward of the reference datum are negative numbers. If the firewall is station 0 (reference datum), then the horizontal stabilizer will have a fairly large positive station number and the engine mount will have a small negative station number.

A wing station is a point on a wing measured right or left from the aircraft centerline, C_L. Where a wing station is given, it is assumed that the reference datum is the aircraft centerline. Wing stations may be addressed as Wing Station Left (WSL) or Wing Station Right (WSR), or by a numbering system where the station lines on the right side of the aircraft are positive numbers, and station lines on the left side of the aircraft are negative numbers.

There may be more stations identified for the horizontal stabilizer (HS), aileron (AS), etc., with those also referenced to the aircraft centerline, C_L.

Buttock Lines

A buttock line is a distance right or left of the aircraft centerline, as viewed from the top. This is the almost the same as wing stations, except buttock lines are used to address locations in the fuselage (usually, they sometimes are used to address locations in the wing and horizontal stabilizer as well). They are abbreviated as Buttock Line Left (BLL) or Buttock Line Right (BLR), or by a numbering system where buttock lines on the right side of the aircraft are positive numbers, and buttock lines on the left side of the aircraft are negative numbers. Buttock Line 0 coincides with the aircraft centerline, C_L.

Water Lines

A water line (W_L) is a vertical distance in the fuselage, as viewed from the side. The 0 Water Line is typically through the fuselage about in the middle as viewed from the side, and parallel to the flight path. It may or may not coincide with the thrust line (T_L) shown on the drawing. Water lines above the zero water line are positive numbers and water lines below are negative numbers.

Airplane Components and Definitions

The followings concepts and terms are used extensively in this manual.

Powerplant

The engine and propeller combined.

Fuselage

The airplane structure *not* including;

- the wings, and parts that attach to the wings,
- the horizontal and vertical stabilizers, and the parts that attach to them,
- the powerplant or the engine mount (where the engine mount is removable from the airplane).

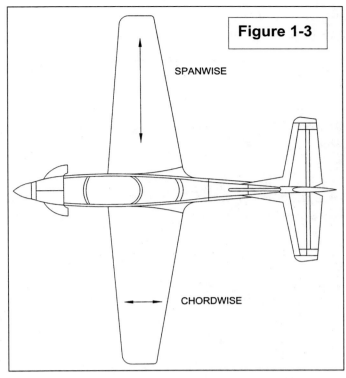

Figure 1-3

Airframe
The whole airplane structure not including the powerplant.

Wing
The entire component designed to produce the majority of lift. In aircraft design it also typically includes the portion of the wing in the fuselage, but for rigging it is that wing area which is exposed to the airflow. The tail components should also be thought of as wings although they will be addressed specifically when discussing them. In discussing wings (including tails), the terminology seen in Figure 1-3 will be encountered frequently. The terms are used loosely to indicate a general direction along the arrows.

Wing Panel
Used to describe the individual sections of a wing which is not a one piece wing.

Wing Section
The airfoil or cross-section of a lifting surface (lifting surfaces include wings, canards, vertical stabilizer/rudder, horizontal stabilizer/elevator, etc.). See Airfoil Geometry in this chapter.

Wing Planform
The shape of a wing or other lifting surface as viewed from above (along the vertical axis). Some examples are given in Figure 1-4.

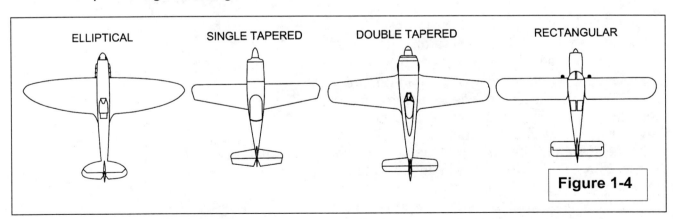

Figure 1-4

Wing Root and Wing Tip
The *wing root* may mean two different things, taken in context;

- the area of the junction of the wing and fuselage, perhaps aerodynamically faired to reduce interference drag, or;
- the physical attach point(s) of the wing to the fuselage (spar fittings).

The *wing tip* may mean two different things, again taken in context;

- the most outboard section of wing, aerodynamically shaped to minimize performance losses from vortices, and generally carries little or no structural loads, or;
- the most outboard rib or other structural piece that defines the wing section, exclusive of a nonstructural tip.

Aileron

That movable portion of the wing that is used to control roll. The combined wing and aileron are considered to be a [flapped] wing in aerodynamics.

Vertical Stabilizer and Rudder

The combined vertical stabilizer and rudder are a [flapped] wing. The vertical stabilizer is the fixed (unmovable) section used to create directional stability. The rudder is the rotating (movable) section that is used to control yaw.

Horizontal Stabilizer and Elevator

The combined horizontal stabilizer and elevator are a [flapped] wing. The horizontal stabilizer is the fixed (unmovable) section used to create longitudinal stability. The elevator is the rotating (movable) section that is used to control pitch.

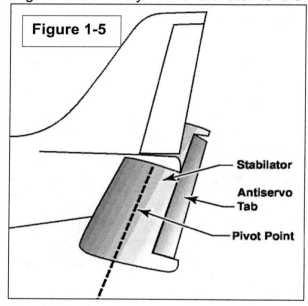

Stabilator

The stabilator is found on some aircraft designs. It is a horizontal stabilizer and elevator combined (Figure 1-5). The entire horizontal stabilizer pivots to produce pitch control while still acting to produce the necessary longitudinal stability for the airplane. Stabilators are advantageous, on some aircraft, because they require less area to provide the same aerodynamic forces. The anti-servo tab in the figure is discussed later.

Trim Tab

A trim tab is the portion of a control surface that is used to force that control surface, aerodynamically, to a desired position (Figure 1-6). They are used to relieve control forces during flight. A few small aircraft may not have any trim tabs, but most aircraft have them on at least the elevator. The *geometric* neutral position of the trim tab is when it is aligned with the control surface that it influences. The neutral setting for elevator trim tabs that is indicated in the cockpit on small planes is; the position that results in the aircraft having desirable control forces during takeoff, at some center of gravity position. This may or may not be streamlined with the elevator, depending on the airplane design. Aircraft with large CG ranges have different takeoff trim settings for different CG locations.

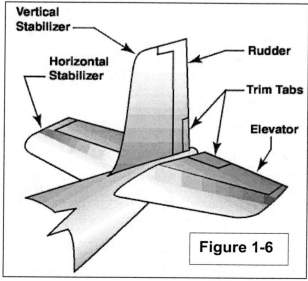

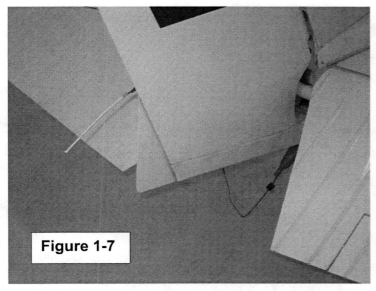

Figure 1-7

Trim tabs may be adjustable from the cockpit or only adjustable on the ground. The elevator trim tab is usually adjustable from the cockpit to allow the pilot to relieve the constantly changing elevator control forces caused by speed and configuration changes of the airplane.

Fixed trim tabs are simply pieces of aluminum attached to the control surface which may be bent on the ground to provide a trim condition for one airspeed (often cruise flight where it is desirable to be able to relax the hands and feet without the airplane going off in an undesired direction). An example of a fixed trim tab is shown on the rudder in Figure 1-7.

NOTE

Force trim is an alternate method to hold the control surface in a desired position by mechanical means. It works by applying a force on the control surface through a spring, whose tension is cockpit adjustable.

A few small airplanes are trimmed longitudinally by adjusting the incidence of the horizontal stabilizer (Figure 1-8) (incidence is discussed later in this chapter).

Servo and Anti-Servo tabs

Servo/anti-servo tabs are not found on all airplanes. A servo tab is a second rotating [movable section] of an elevator that deflects in the opposite direction of the elevator to assist the elevator in movement (Figure 1-9). It deflects automatically as the elevator is moved and is used to reduce the pilot effort necessary to deflect the elevator.

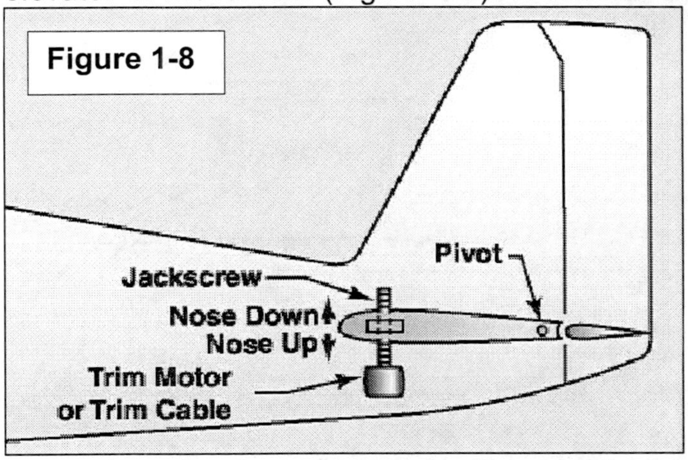

Figure 1-8

An anti-servo tab is a second rotating (movable section) of an elevator that deflects in the same direction as the elevator. It deflects automatically as the elevator is moved and is used to increase the pilot effort required to deflect the elevator. It is commonly used on stabilators to provide some stability to the 'floating' horizontal stabilizer (Figure 1-5).

Servo or anti-servo tabs may also function as a trim tab when equipped with a method of changing their angle separately from the elevator.

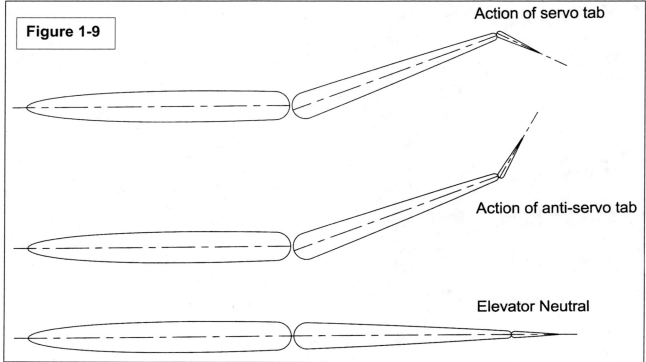

Figure 1-9

Action of servo tab

Action of anti-servo tab

Elevator Neutral

Spoilers

Spoilers are not common on small airplanes but are frequently used on gliders (Figure 1-10). Spoilers may be used to control roll when deployed differentially, or to lose lift when deployed simultaneously. Spoiler installations are generally designed to do one or the other, and are commonly used on gliders to lose lift. When used for roll control, there is a slight delay between the spoiler deployment and the aircraft reaction, making them undesirable for most small airplanes.

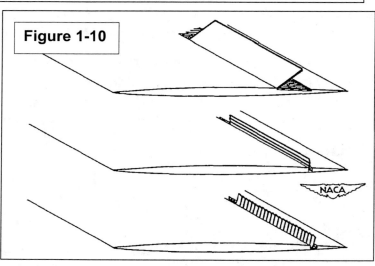

Figure 1-10

Empennage

A term used to describe the portion of the airplane that includes the vertical stabilizer, rudder, horizontal stabilizer, and elevator, collectively.

Control Surface

The control surfaces are the ailerons, elevator, rudder, flaps, trim tabs, spoilers, etc.; any part capable of movement independently of the airframe, that is used to aerodynamically control aircraft attitude or performance.

Biplane Parts

Refer to Figure 1-11 for common terminology of the various parts used mainly on biplanes. The most common difference between various biplanes, structurally, are the arrangement of the cabane struts and wires that support the center section of the upper wing, and the interplane struts. Also the upper wing is commonly found to be one or three pieces, sometimes two.

Cabane wires, where present, are usually used in lieu of some of the cabane struts, and are oriented mainly fore and aft. They are sometimes referred to as stagger wires, or drag/anti-drag wires.

The roll or transverse wires (if present on a particular design) are located in the cabane framework and are mainly oriented spanwise.

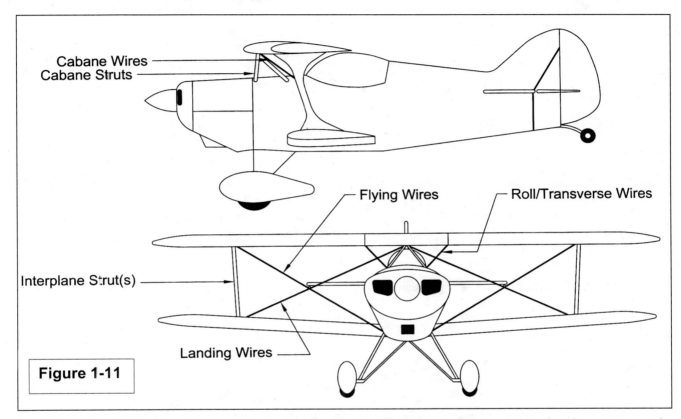

Figure 1-11

In discussing biplanes, the *cellule* is the arrangement of both wings, the 'box' so to speak. When referring to the structure and loads of the wings, the cellule includes the wings, wires, and struts. When discussing biplane aerodynamics, the cellule is the total of the aerodynamic force vectors that the wing combination produces (one wing affects the other) and is generally considered to only include the two wings (no struts, wires, etc. since their overall effect is small).

Airfoil Geometry

An airfoil is also referred to as a *wing section* (note that the empennage is also composed of wing sections). An airfoil may be symmetrical or asymmetrical.

Refer to Figure 1-14 for the following discussion

Chord Line

The *airfoil chord line* is a straight line that extends from the leading edge to the trailing edge.

NOTE

The exact point which is considered the leading edge of the airfoil (for the purposes of defining the chord line) is defined mathematically for a particular airfoil, and is not always the most forward part of the wing section. It is usually close enough though that rigging instructions for airplanes will reference the most forward part of the wing as the leading edge to simplify the work for the rigger.

Angle of attack measurements are made in reference to the airfoil chord line. Angle of attack is defined here rather than Chapter 2 because the idea is used so frequently in discussing aircraft geometry. The angle of attack is the angle between the relative wind and the chord line of the airfoil (Figure 1-12). It is positive when the relative wind comes from underneath the chord line, and negative when the relative wind approaches the chord line from above. The greater the angle of attack, the more lift is produced (up to a point, it is discussed in more detail in Chapter 2).

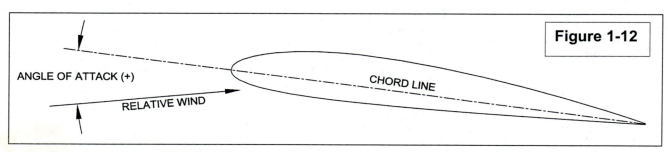

Figure 1-12

In rigging, reference may also made to the *wing chord line*. A wing chord line is parallel to the centerline of the aircraft, like WSR 56 or WSL 56 in Figure 1-2. The airfoil chord line follows this line, even when the wing leading and trailing edges are swept (Figure 1-13). This is important in determining the control deflections (discussed later).

NOTE

Most wings of small airplanes have the structural ribs similarly arranged, that is they are shaped like the airfoil used and are arranged parallel to the longitudinal axis of the aircraft. Some airplanes with delta or swept wings have the ribs at an angle to the spar(s) for structural reasons, and those ribs are not the same shape as the airfoil.

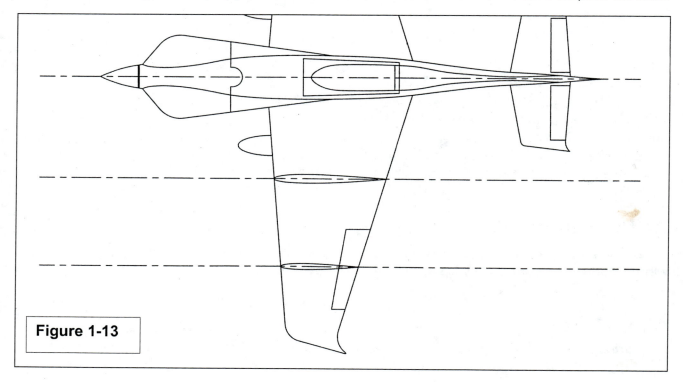

Figure 1-13

Mean Camber Line

The *mean camber line* divides the upper and lower halves of the airfoil into equal parts. The camber (amount of curvature) is the distance between the chord line and mean camber line at its greatest point, and is a major indication of the aerodynamic properties of an airfoil.

Symmetrical Airfoil

A symmetrical airfoil has the same camber on the top and bottom (upper half of Figure 1-14). Because it is the same on the top and the bottom, the camber line and chord line are the same (the camber line is straight).

NOTE

On most aircraft, the vertical and horizontal stabilizers are symmetrical airfoils.

Asymmetrical Airfoil

An asymmetrical airfoil has more camber on one side (illustrated in the center and bottom of Figure 1-14).

Zero Lift Line

The zero lift line is used to illustrate the direction of the air striking the airfoil when the airfoil produces zero lift. On a symmetrical airfoil, the chord line and zero lift line are the same. On an asymmetrical airfoil, the zero lift line is determined experimentally in a wind tunnel, but can be approximated as a function of the camber. The zero lift line (degrees) is approximately equal to the camber (percent) of the airfoil. For example, an airfoil with 2% camber has a zero lift line approximately 2° different than the chord line (the bottom of Figure 1-14). For the cambered (asymmetrical) airfoil, the angle of attack must be negative to produce no lift.

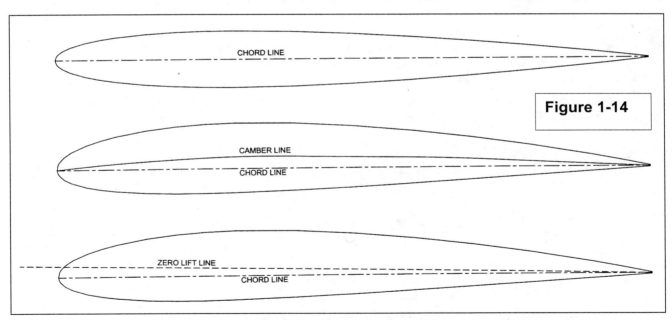

Figure 1-14

The zero lift line of a wing will be different than that of the airfoil, and is determined by the wing geometry (planform, aspect ratio, washout, etc., discussed again in Chapter 2).

Incidence Angle

Incidence angle is defined as the angle between the chord line of the airfoil and the airplanes' longitudinal axis (Figure 1-15).

Incidence angle is a positive number when the trailing edge of the chord line is further down than the leading edge of the chord line, in relation to the airplane longitudinal axis (the airfoil is rotated nose-up), and a negative number when the trailing edge of the chord line is further up than the leading edge of the chord line (rotated nose-down).

On some airplanes, the vertical stabilizer is set with incidence to compensate for aerodynamic phenomenon caused by the propeller (discussed later). The vertical stabilizer incidence is measured in relation to the aircraft centerline.

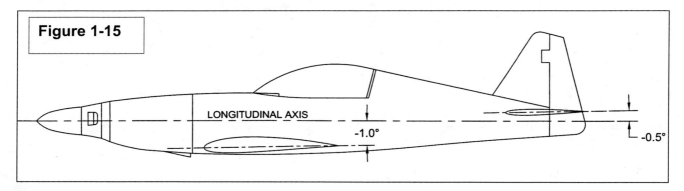

Figure 1-15

Wash-out

Wash-out is the *geometric* twist of the wing (found on some airplanes), such that the inner portion of the wing is at a slightly higher incidence angle than the outer portion (Figure 1-16). Wash-in is the opposite, the wing tips are at a higher incidence than at the wing root. Wash-out is assigned a negative number and wash-in is assigned a positive number. The reasons for wing twist are discussed in Chapter 2.

NOTE

For the purposes of clarity, the illustration in Figure 1-16 only shows the difference in incidence between the wing root and wing tip, rather than the actual incidence of the wing sections compared to the longitudinal axis of the airplane.

Related to this is *aerodynamic washout*, where a different airfoil is used on the outboard section of wing. This produces aerodynamic effects that are similar to geometric washout.

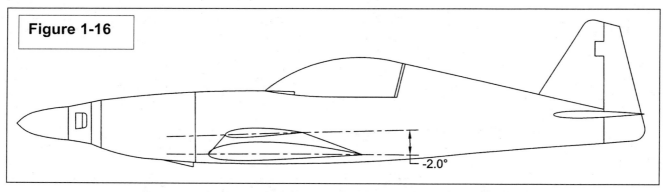

Figure 1-16

When a wing is twisted to provide washout, the specified incidence angle of the wing in relation the airplanes' longitudinal axis is measured from a specific wing station, since the incidence would be different at every location on the span of the wing.

Thrust Line

The thrust line is the orientation of propeller thrust with regards to the longitudinal axis of the airplane. For the purposes of rigging it is considered to be perpendicular to the propeller plane of rotation (propeller disc) with the origin at the center of the propeller hub (Figure 1-17). The thrust line may be aligned with the longitudinal axis of the airplane, however, certain airplanes are designed with the thrust line off at an angle (Chapter 2).

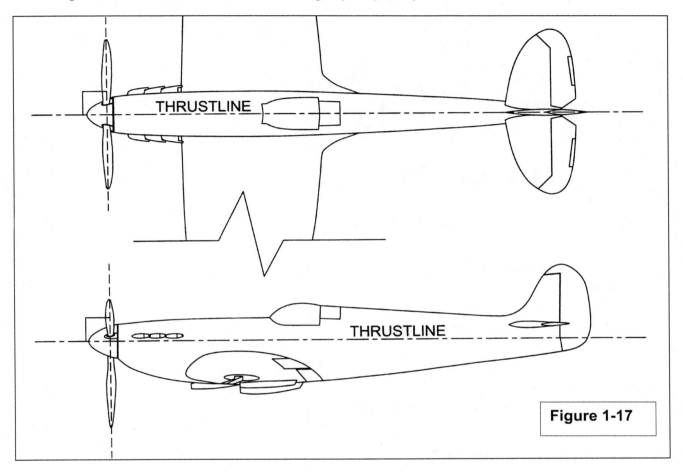

Figure 1-17

Dihedral

Dihedral is the term used to describe the geometry where the wing tips are higher than the wing roots. Anhedral is the opposite, the wing tips are lower than the wing roots. It is the angle that the wings make with the lateral axis of the airplane (Figure 1-18). It is used to provide 'lateral stability'. Many aerobatic aircraft don't have any dihedral because *stability opposes control*.

In terms of airplane layout, the dihedral is measured from the *mean chord plane*, which is a plane that passes through the chord line of the airfoil(s) that make up the wing section. Because many wings taper in thickness from root to tip, the angle measured on the top or bottom of the wing will be different than the actual dihedral angle. Most designers will specify a dihedral angle in relation to the top or bottom of the wing to make rigging easier.

If the wing has washout, or uses different airfoils between the root and tip (see Washout in this chapter), the mean chord plane is a *line* that passes through the chord line of all the wing sections. Because this will occur at only one chordwise location, that chordwise position becomes important to taking the dihedral measurement, and designers may specify a dihedral angle that is to be measured at a particular chordwise location.

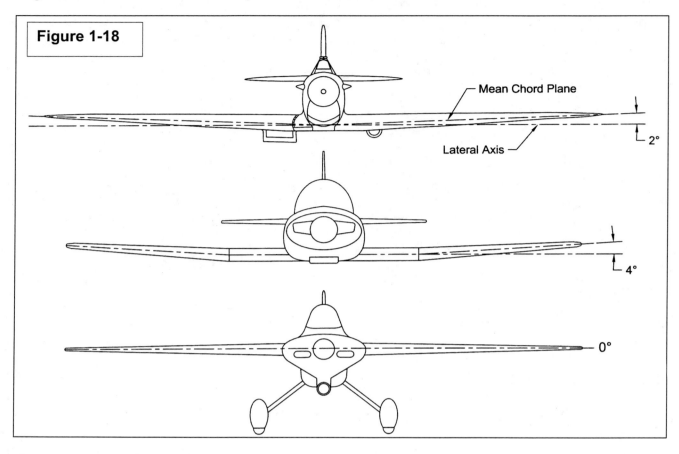

Figure 1-18

Control Deflection

Control deflection is the angular rotation that a control surface/flap may be moved (Figure 1-19).

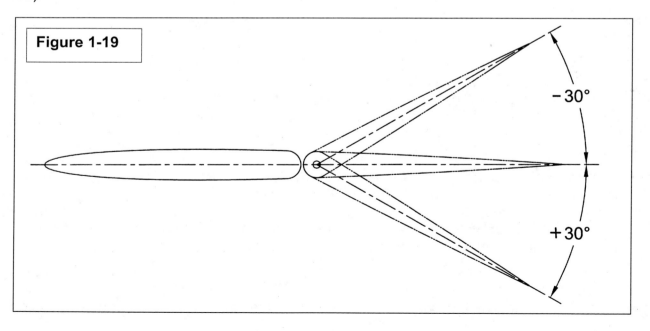

Figure 1-19

A measurement given as a negative deflection means that the deflection is trailing-edge-up; positive deflections are trailing-edge-down. The angle is made in reference to the chord line of the airfoil, however it doesn't matter for the purposes of rigging because if the measuring tool is always placed either on the top or bottom of the control surface, the same angle will result (setting the neutral position of the control surface requires an extra step, discussed in Chapters 3 and 5).

Angular deflection measurements are made parallel to the wing chord line, unless it is specifically stated that they be made perpendicular to the hinge axis (Figure 1-20).

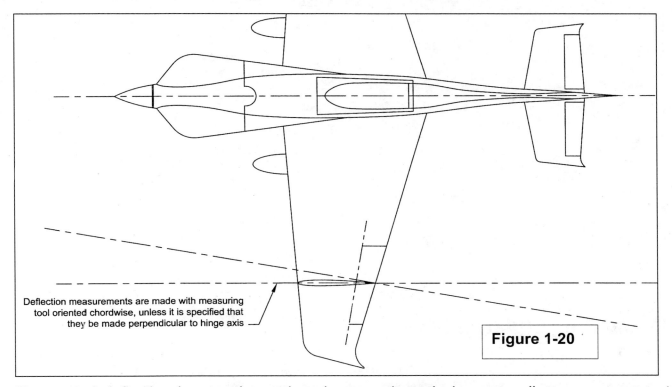

Deflection measurements are made with measuring tool oriented chordwise, unless it is specified that they be made perpendicular to hinge axis

Figure 1-20

The control deflection is sometimes given in manuals and plans as a linear measurement made from the trailing edge of the control surface as in Figure 1-21 (see how to convert between the two in Appendix A). When a control surface varies in chord length along the span like the aileron in Figure 1-20 (it tapers), a linear measurement of deflection is only valid at a single spanwise station of the control surface. This is illustrated in Figure 1-21.

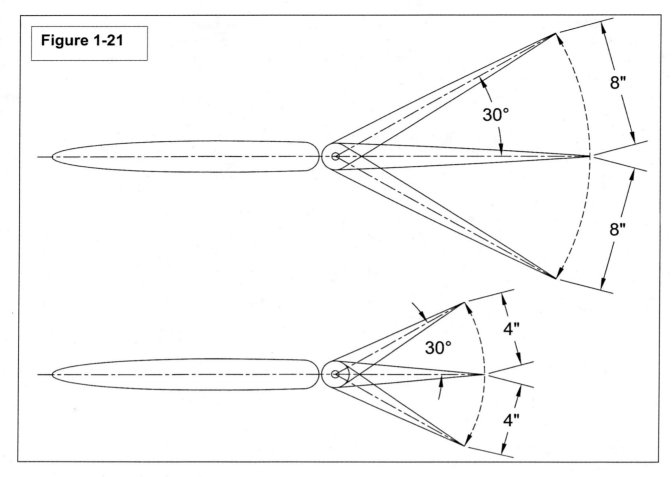

Figure 1-21

Decalage (Biplanes)

Decalage (Figure 1-22) is the *difference* in incidence angle between the two wings of a biplane (see incidence angle earlier in this chapter). It is a positive number when the zero-lift lines of the wings intersect ahead of the airplane, and a negative number when they intersect behind the airplane. For the purposes of rigging, the measurements are usually taken from the chord lines of the wing sections, rather than the zero-lift lines, to simplify the construction of the rigging tools. Therefore decalage may be taken to mean two different things depending on whether design or rigging is being discussed (except for symmetrical airfoils whose zero-lift line coincides with the chord line).

Stagger (Biplanes)

Stagger is the difference in longitudinal location of the two wings of a biplane. For the purposes of rigging, it is typically referenced to some point on the leading edges to simplify measurements (right side of Figure 1-23).

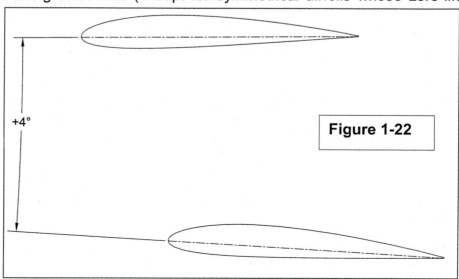

Figure 1-22

Because the upper and lower wings may be different shapes, in airplane design the measurement of stagger is between some fixed location on the mean aerodynamic chord of both wings, usually 25% aft of the leading edge of the MAC, denoted C/4 in the figure (MAC is discussed later in this Chapter).

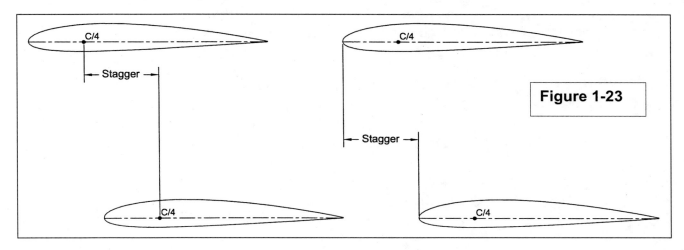

Figure 1-23

Gap (Biplanes)

Gap is the vertical distance between the two wings of a biplane (left side of Figure 1-24). Because the two wings of a biplane usually have some decalage, the gap is measured at the 25% MAC point (MAC is discussed later in this Chapter) for the purpose of aircraft design. Airplane rigging cannot alter the gap substantially.

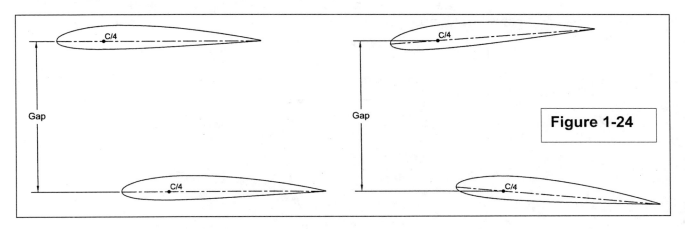

Figure 1-24

Mean Aerodynamic Chord

Although the Mean Aerodynamic Chord (MAC for short) is more of an aspect of airplane design, the concept is used extensively in aircraft layout and weight and balance. The significance of the MAC is provided in this section, however, most of the material here will not be encountered again. It is important to have an idea of what the MAC is since the term is commonly used.

The MAC is a method of reducing different shapes of wings to a simple rectangular layout for the purposes of positioning the wing during design, and performing stability calculations. That simple rectangular layout is the *geometric average* of the wing planform in question. It is also used in weight and balance computations by the aircraft operators (the MAC isn't often used for weight and balance of small airplanes but the term may be used in airplane drawings for layout).

The MAC is a *wing chord line* (see the previous paragraph on airfoil geometry) derived from the geometric average of the whole wing. It accurately reflects the aerodynamic pitching moments (Chapter 2) produced by a complex wing shape, by reducing them to an average wing chord (hence the name MAC). An example is given;

The fictitious rectangular wing that the MAC is derived from is drawn accurately in Figure 1-25 for the swept wing airplane. It is the geometric average of the whole swept wing.

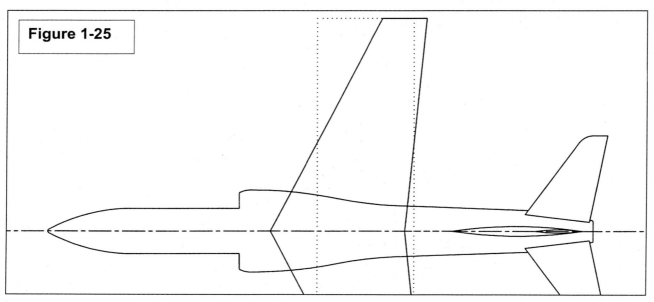

Figure 1-25

The layout lines in Figure 1-26 are the method of graphically determining the geometric average of the wing planform (instead of by computation). It isn't normally necessary to know how to do this.

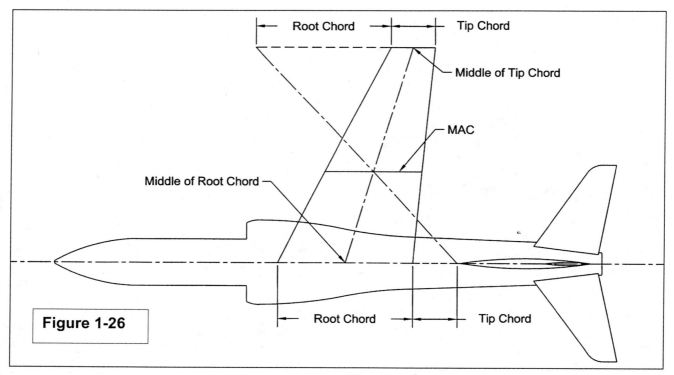

Figure 1-26

When the MAC is made known, the fictitious wing is illustrated as surrounding the MAC (Figure 1-27). The fictitious wing is only here to illustrate the concept. The important part is the MAC itself, and at what fuselage stations are the leading and trailing edge of the MAC (abbreviated LEMAC and TEMAC). Commonly referred to for layout is the quarter chord point of the MAC (C/4 in Figure 1-27). This is approximately the aerodynamic center of the wing, where the aerodynamic pitching moments produced by the wing remain constant with changes in angle of attack. It is 25% of the MAC aft of the LEMAC, so for this example would be FS 244.25.

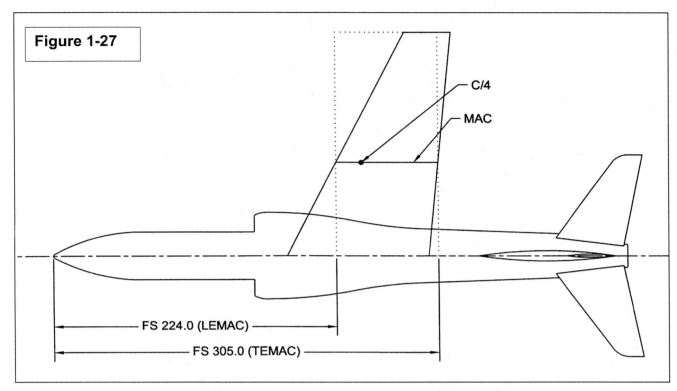

Determining the MAC for other wing planforms like double taper and elliptical is more complicated, but can still be done graphically or computationally. The airplane drawings, if they reference the MAC, will already provide the fuselage stations for the LEMAC and TEMAC.

Chapter 2
Aerodynamics and Flight Mechanics Affecting Rigging

It is important to have an understanding of aerodynamics and flight mechanics to effectively (and safely) experiment with the rigging of an airplane. The information given in this section is very general, but will provide some insight to rigging requirements and provide an understanding of the factors involved. This chapter assumes some knowledge of aerodynamics, a good starting point is Reference 22. Some subjects are repeated. Because some aspects can best be described mathematically, a little algebra and some graphs are given to illustrate the important points.

It is difficult to separate airplane rigging from airplane design, as rigging is dependent on design. This manual attempts to avoid discussing design and concentrate on those aspects which may be altered easily by the builder or mechanic, however, it is necessary sometimes to discuss the science that drives the mechanics. Altering an airplanes' geometry can produce unexpected and potentially dangerous results if not approached in a methodical nature.

This section discusses a number of different concepts. Some of them are necessary to introduce other subjects, while others may be taken alone as having an influence on the rigging procedure.

Rigging problems can be difficult to diagnose because of the aerodynamic interaction between the flight axes of the airplane. Sometimes it is difficult to describe what the airplane is doing and some concepts are taken from the study of flight mechanics to provide the rigger and pilot with unambiguous language.

Because of the wide variety of airplane shapes and configurations, it is only possible to make generalized statements and inferences about airplane behavior. It is up to the rigger to ensure that rigging changes are within the specified limits or that the airplane is carefully test flown. As is the case with other aspects of the science, there are always exceptions and unforeseen consequences. Sometimes it takes experimentation to achieve the desired result. Airflow around a body can separate and attach very quickly, sometimes causing small rigging changes to have drastic effects, positive and negative. Since these phenomenon are specific to the shape of the body and the particular phase of flight, these things are generally only found through flight testing or wind tunnel testing.

Vectors

A brief discussion of vector notation is necessary because the concept and associated terminology is used frequently in discussing the various forces imposed on an airplane. Nowhere in this manual will there be any mathematical solving of vectors, but it is important to understand the concept before proceeding because vectors are used frequently to describe various effects.

A vector is used to graphically describe a force or velocity. It is symbolized by a line with an arrowhead. The direction of the vector coincides with the direction of the force/velocity, and the magnitude of the force/velocity is given by the length of the line (Figure 2-1). Vectors may be solved mathematically, or, by drawing them on a piece of paper and solving them graphically.

In Figure 2-1, a force is being applied at a 45° angle with a force of 5 units. To find the amount of force that will be acting along 0° or 90°, see Figure 2-2.

In the case of two forces acting at the same time, the solution is had by the addition of the vectors as in Figure 2-3. This example uses two forces which are 10 and 5, acting on an angle of 45° and 330° respectively. The resultant force has a magnitude of 12 and is acting on an angle of 22°. More than two forces may be solved for in this manner by continuously adding more vectors on to each other in the manner of Figure 2-3. Note that the vectors may be made into triangles and solved with trigonometry instead of drawing them out.

The vectors illustrated in the previous examples are two-dimensional but the forces acting on an airplane are three-dimensional. The concept is the same but the graph gets a third dimension (the Z axis in Figure 2-4).

The vectors acting on an airplane may be labeled lift, weight, thrust, and drag. Each individual component of an airplane produces its own lift and drag (and has weight), and each section of an individual component produces a different reaction than the adjacent part. Weight doesn't always act towards gravity in an airplane, it is the vector sum of the gravitational force and the centrifugal forces of maneuvering. All of these aerodynamic and mass vectors are too many to draw all at once; it can approach infinity depending on the detail of the analysis. But, the vector sum of all of these forces of all the individual parts of the airplane determines the

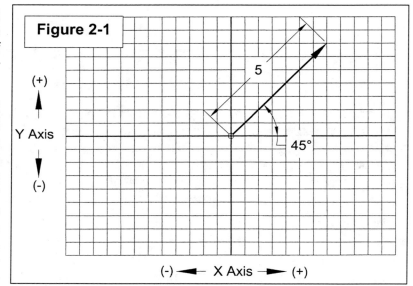

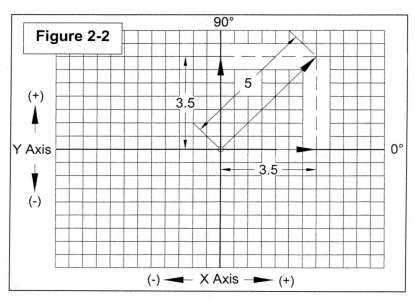

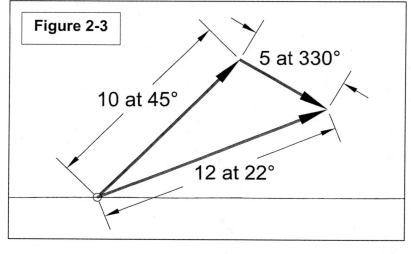

attitude that the airplane stabilizes in flight, or, determines the direction of rotation that the airplane is changing in. When the airplane is moving, those forces will also change their magnitude and direction depending on the speed and power setting of the airplane, and the airplanes orientation to the relative wind at any given instant. It is important to think about the airplanes actions in terms of its various vectors to help isolate the effects of the aerodynamic and mass loads acting on it.

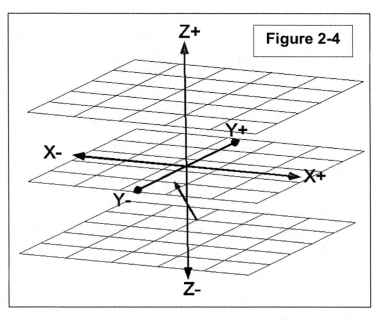

Some vectors are shown on the research aircraft in Figure 2-5 that is rolling and turning (this aircraft rolls by using differential elevator deflection). In this case, the simulator is calculating the aerodynamic force of the lifting surfaces by dividing them up into small chordwise strips and calculating the forces at each little strip. The lift and drag vectors of the flying surfaces in the illustration are represented in additive form (the little lines on top of the long ones). There is also one very long thrust vector coming out of the nose cone and one very long parasite drag vector hidden in the exhaust plume.

Aircraft Movement

An aircraft has six degrees of freedom of movement. Three degrees of freedom are *rotational* and the other three are *translational*, discussed separately in subsequent paragraphs. Rigging (and also flying, for the novice) is made complicated by the fact that the six degrees of freedom are directed with only three controls (four if the throttle is included), and a change about one degree of freedom often results in changes about one or more other freedoms simultaneously.

Rotation

The three rotational degrees of freedom are pitch, roll, and yaw (Figure 2-6). An aircraft rotates about its' center of gravity. The center of gravity of an aircraft is a three dimensional point in the fuselage somewhere. Figure 1-1 illustrates the aircraft axes for layout and the same axes are used to describe the flight attitude of the aircraft. When considering the flight axes, each axis passes through the center of gravity of the aircraft. The airplane controls (rudder, ailerons, and elevator) produce rotation around the CG.

Pitch is rotational motion about the lateral axis (the longitudinal and vertical axes rotate about the lateral axis).

Roll is rotational motion around the longitudinal axis (the lateral and vertical axes rotate about the longitudinal axis).

Yaw is rotational motion about the vertical axis (the longitudinal and lateral axes rotate about the vertical axis).

The axes illustrated here are the *body* axes and are discussed again later.

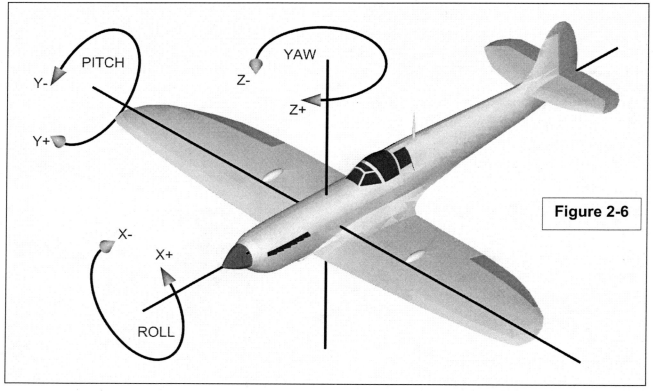

Figure 2-6

Translation

Translation is movement along a line, or, 'in a direction' (Figure 2-7). It can be symbolized by a vector whose magnitude (length of the line) represents velocity. An airplane may be translated longitudinally (forward/backwards), vertically (upwards/downwards), and laterally (left/right), by some combination of control inputs. An airplane translates forward as its normal mode of flight and the vector sum of the translations are predominately forward.

The vector addition of the three lines of translation result in a three dimensional flight path of the airplane. An aircraft may be traveling in a straight line even though it is pointed in a different direction. The concept of translation is important to understanding airplane rigging because an airplane that is out of rig will translate through the air in an undesired manner. In rigging, the three translational modes may be used to compare the direction that the airplane is moving in relation to the orientation of the body axes, and geometry adjustments made to more favorably orient the body axes to the translational directions. An airplane produces lift and drag forces in many directions simultaneously (and sometimes asymmetrically), and the vector sum of those forces describe a resultant three dimensional flight path of the airplane and its' attitude. Airplanes with propellers tend to translate undesirably because of propeller forces and is discussed in more detail later.

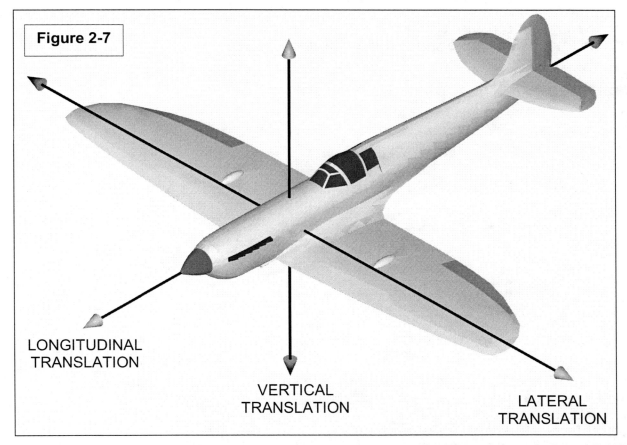

Figure 2-7

Most translations of an airplane occur in combination with forward translation. Backwards translation isn't too common except in tail slides (with reference to the airflow) or flying in a very strong headwind (with reference to the earth).

Vertical translation is usually a result of changing the lift of the wing (note that an airplane does not climb or descend because of changes in lift, see Reference 22). A good example of translation on the vertical axis is pointing the airplane straight down and accelerating at a constant pitch angle (90°) as in Figure 2-8. As speed increases, lift increases and causes the airplane to translate along its' vertical axis while the longitudinal axis remains at a fixed angle to the earth.

Lateral translation generally occurs when roll and yaw control inputs are opposite each other. This is most easily demonstrated in forward slips where the longitudinal axis of the airplane is pointed in a different direction than the airflow, and a different direction from its' travel over the earth. A side slip like the kind used to land in a crosswind is also a form of lateral translation, with the longitudinal axis of the airplane being pointed in a different direction than the airflow, but lined up with the direction of travel.

One of the most common rigging problems is the condition where the airplane is continually in a slight slip (see slips and skids for definitions).

Straight and Level Flight

It is important to define straight and level flight. Straight flight is traveling in a constant direction, or heading. Level flight is traveling at a constant altitude. In either case, the aircraft won't necessarily be pointed in the direction it is moving.

Accelerated and Unaccelerated Flight

Taken in context, acceleration is either changing speed or changing direction (or both). Unaccelerated flight is not changing direction or speed, in equilibrium at 1 g (gravity).

Aircraft Coordinate Systems

Describing aircraft movement is further complicated by the fact that it is possible to describe the aircrafts movement in terms of;

- it's flightpath relative to the flight axes,
- it's flightpath relative to the oncoming air (relative wind),
- its' flightpath relative to the earth.

Airplane rigging is mainly concerned with the difference between the first two, flight axes and relative wind, but because we use the earth as a reference, it can be an important clue as to what the airplane is doing (in a no wind condition).

Body-Axis Coordinate System

The body-axis coordinate system remains coincidental to the aircraft; it moves with the aircraft (Figure 2-9). The various axes all intersect at the center of gravity of the aircraft (the origin of this coordinate system) and aircraft rotation (pitch, roll, and yaw) occur about the body axes. This axis system is also commonly referred to as the flight axes.

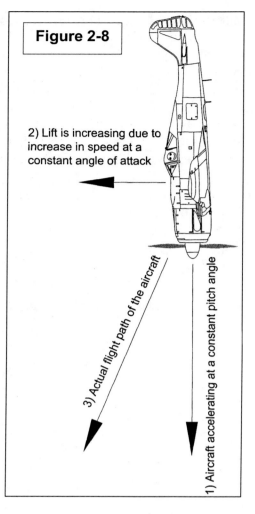

Figure 2-8

2) Lift is increasing due to increase in speed at a constant angle of attack

3) Actual flight path of the aircraft

1) Aircraft accelerating at a constant pitch angle

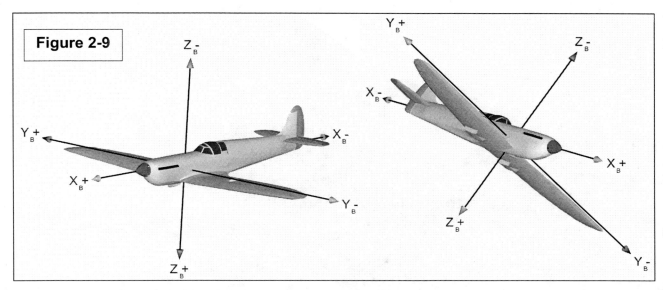

Figure 2-9

Wind-Axis Coordinate System

The wind axis system is used to describe the direction of the relative wind (Figure 2-10). Mainly it is thought about in terms of the wind vector approaching the front of the aircraft, along X_W. The origin of this axis system is typically the center of gravity of the aircraft (but could be anywhere convenient). In Figure 2-10, the left aircraft is in a slow descent and X_W is coming up from underneath. The right aircraft is rolled and yawed in opposite directions (slipping) and the relative wind (X_W) is striking the left side of the aircraft.

Much airplane rigging is concerned with keeping X_W and X_B coincidental. An aircraft typically has the least amount of drag when X_B is the same as X_W. Obviously the angle of attack is always changing as a consequence of flight, but performance will be greatly improved if at least X_W is aligned in the plane of symmetry along X_B (Figure 2-11).

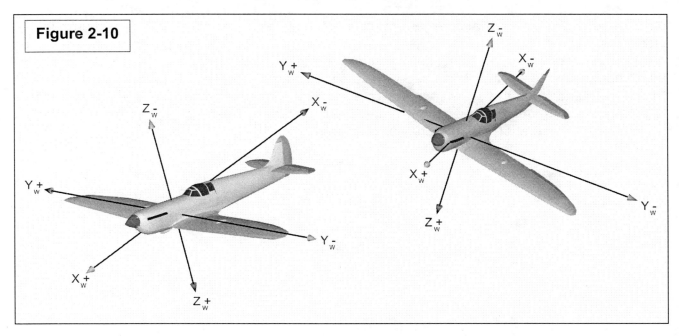

Figure 2-10

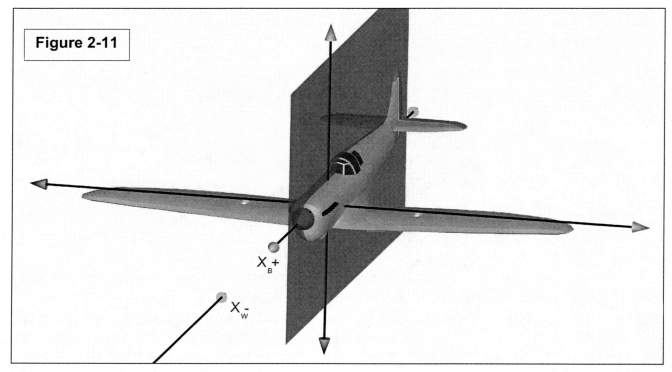

Figure 2-11

The relative wind is the X_W axis. For the purposes of rigging, it is important to not think about it only in terms of the wing's angle of attack as it strikes the various parts of the aircraft. In addition, airflow is altered as it passes around the aircraft, striking the aft sections of the aircraft at some different angle (and velocity). This is referred to as the *local relative wind* or *local airflow*.

Earth-Axis Coordinate System

The earth-fixed coordinate system remains oriented to the center of the earth (Figure 2-12), regardless of the aircrafts attitude or direction. The Z_E axis acts along gravity, X_E is oriented North-South, and Y_E is oriented East-West. The origin is on the surface of the earth. The earth is assumed to be flat and not rotating. Although this coordinate system doesn't have much application to rigging, it helps to visualize the flightpath of the aircraft since the earth and the aircraft tend to be the pilots frame of reference for aircraft control. This coordinate system is mainly used in stability and control analysis to provide a reference frame for the effect of gravitational acceleration on the behavior of the aircraft.

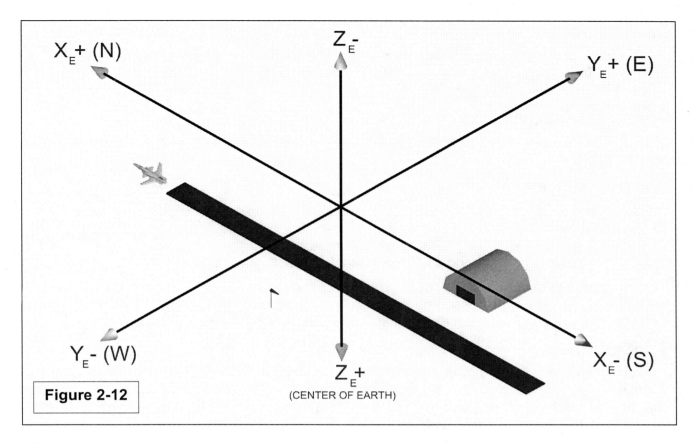

Figure 2-12

Moment

A moment is the magnitude of a force that rotates or tries to rotate an object around a point. It is a measurement of torque. It may be produced by either aerodynamic reaction, mass, or both. In aerodynamics, pitching, rolling, and yawing moments are the forces that rotate the aircraft.

At a given airspeed and angle of attack, a given amount of control deflection will produce a given amount of aerodynamic force. That aerodynamic force, multiplied by its distance from the CG of the aircraft, is the moment produced that rotates the aircraft about the CG. It will continue to rotate the aircraft until the moment is cancelled out by other moments, or a change in speed, angle of attack, or thrust changes the original force.

Hinge moments are the forces that try to rotate the control surfaces (back to neutral normally). They are usually due to the airloads on them, but inertia may also attempt to deflect the controls in certain situations (maneuvering or fluttering).

Mass moments are the distribution of weight in relation to the CG, used in weight and balance calculations, and in stability and control analysis.

Aircraft Control

Aircraft control is the ability to maintain or change the attitude and/or flight path to that which is desired. The rigger is mainly concerned with the ability of the controls to produce the desired result, but some rigging can affect the fundamental controllability and stability of the aircraft. Control can be described in several aspects, and different textbooks use different definitions.

Control Force

Control force is the amount of pressure required at the control actuator to produce the desired change or rate of change in the aircraft attitude, or, the amount of pressure necessary to deflect the controls a certain amount.

Control Effectiveness

Control effectiveness is a subjective (unitless) evaluation of a controls ability to produce the desired result.

Control Response

Control response is a subjective (unitless) evaluation that indicates that the aircraft does what is expected or desired when the control is deflected. For example, it may be termed as 'fast' or 'sluggish', or that it is sluggish at first, then becomes fast, or vice-versa. It may be used to say that the aircraft rolls in the wrong direction when the ailerons are applied.

It is sometimes quantified as; the rate of rotation of the aircraft as a function of control surface deflection. It is useful for describing the response of the aircraft to control inputs if they are very non-linear (typically an aerodynamic problem but may be the control linkage design).

Control Power

Control power is the force that is available from a control surface in a given flight condition, to produce an angular acceleration of the aircraft about its CG. For example, an aircraft flying at slow speeds will have decreased control power for a given deflection. It may be expressed as a quantitative value of pounds of force acting normal (perpendicular) to the control surface which in turn produces an aircraft rotation rate (degrees per second) or acceleration (degrees per second per second).

Control Authority

Control authority is the amount of *control power* available in a given flight condition. It changes with airspeed, center of gravity, angle of attack, and the direction of the relative wind (X_W).

For example, a propeller aircraft climbing steeply requires much rudder to keep the aircraft going straight, reducing the amount of rudder available for a turn in one direction; hence, the control authority of the rudder is reduced in that direction. Another example which is common is the conduct of a sideslip during a crosswind takeoff or landing. Because the rudder and ailerons are deflected to perform a sideslip, there is less control available to produce roll or yaw in the direction the controls are deflected.

Excessive control forces can also have an effect on control authority since only a finite amount of effort may be imposed by the pilot on the controls in a given flight condition.

Control Sensitivity

Control sensitivity is used to describe the amount of control power produced for a given linear displacement of the control actuator. It is often used to describe a situation where very little forces are required on the actuator to initiate a large change in aircraft attitude.

Rigging generally can't alter this for normal aerodynamic controls. The amount of deflection of a control surface for a given amount of actuator movement, and the force required at the actuator to produce that deflection, is mostly a function of the gearing ratio between the actuator and the control being displaced, airloads remaining constant. It is also affected by aerodynamic balancing of the controls (a design issue). The shape of the trailing edge of a control surface has a powerful influence on the feel of that control (and is a method of aerodynamic balancing).

On some control systems, multiple mounting holes are supplied for pushrods, bellcranks, and idlers, in order that the rigger be able to alter the gearing ratio between the stick and the surface that it actuates. In this case it is *usually* desirable that the control actuators have the maximum amount of movement available (with a pilot sitting in the cockpit, the legs often set the limit of deflection of the stick). Making the stick longer can reduce linear sensitivity but produces lighter control forces because of the extra leverage one is able to impose on the controls.

Controls labeled as 'twitchy' are usually taken to mean that they're sensitive and/or have a relatively high breakout force from friction.

On some aircraft there is very little change in stick force required as the control actuator (stick or pedals) is deflected over the whole range. This is undesirable since the change in control force is an important feedback to the pilot. This is usually eliminated from certificated aircraft during the design process, but some homebuilts suffer from this on at least one control axis. It is a design issue (non-linear control linkages or aerodynamic balancing) that causes some aircraft to be labeled as having high control sensitivity at faster speeds.

Trim

To be in trim is the state of the aircraft when it is not changing speed or direction. It is a state of equilibrium, to be in unaccelerated flight. Basic texts will say that when an aircraft is in equilibrium, thrust is equal to drag and lift is equal to weight. It is a little more complicated because thrust helps to overcome weight when the aircraft is pitched up in relation to the earth-axes, and weight also acts on the aircraft during sideslips when the wings are at angle to the earth-axes.

For the purposes of rigging, a state of equilibrium is when the sum of all the force vectors produced by the various parts of the aircraft cancel each other out (add up to zero), and the aircraft remains in a steady state condition (an aircraft with a serious rigging problem may never achieve equilibrium). An airplane in trim doesn't mean that trim tabs are involved or even present, it just means that the vector sum of the forces acting on the aircraft are cancelling each other out (adding up to zero), and the aircraft continues in a straight line at a constant speed. The steady state attitude of the aircraft does not necessarily coincide with the direction of travel.

The concept of trim is important because to produce a state of trim requires aerodynamic forces (lift and drag) that detract from the performance of the aircraft. Those aerodynamic forces are called *trim drag*. Proper rigging can reduce or eliminate some of the aerodynamic forces required to keep an aircraft in equilibrium, but some sacrifices are made for stability.

Take the case of an airplane that is rigged with one wing, or flap, at a slightly different angle than the one on the other side. A slow rolling moment is produced that requires aileron deflection to cancel out. The aircraft is now in a state of trim, whether trim tabs, control friction, or the pilot holds the stick. Because of added drag on one wing, the aircraft yaws slightly and flies slightly sideways (translates laterally), perhaps changing direction very, very slowly. The rudder is depressed to center the ball and align the fuselage with the relative wind. The aircraft is again in a state of trim, but with two controls deflected. The aircraft is now translating sideways slightly and has to be pointed in a different direction to achieve the desired flightpath. The effects are cumulative; there is an increase in drag due to the deflected controls and the control authority in one direction is reduced on those controls. In addition, lift is being used to stabilize the aircraft where otherwise it might be contributing to more performance, and if the aircraft is translating slightly, the airflow around the controls is altered, perhaps increasing the need for more control deflection to establish equilibrium. The thrust vector may be pointed in a different direction than the flightpath, reducing the amount of thrust in the desired direction. This occurs sometimes without the pilot noticing because the attitudes and control deflections involved are small and difficult to detect (there is still a performance penalty though).

Trim and Stability

Stability is related to the state of trim. It is the tendency of an aircraft to return to a state of equilibrium when disturbed, a required characteristic. Stability, by definition, implies an oscillatory movement. Pitch and yaw are the two axes that are subject to oscillatory movement by themselves (longitudinal and directional stability).

Stability opposes control when maneuvering and aircraft are compromised to provide both some stability and some maneuverability. The amount of compromise is dictated by what the aircraft is to be used for (transportation or aerobatics for example).

Lateral stability is not oscillatory by itself. It is coupled with directional stability in its attempt to keep the wings level, termed *roll with sideslip*. Dihedral (Chapter 1) is used to provide roll with sideslip. There are rule of thumb guidelines for determining the amount of dihedral that should be used for a particular aircraft configuration. In general, increasing dihedral without changing the tail size will make the aircraft tend towards free directional oscillations (dutch roll) which is very objectionable (rolling and yawing in the opposite directions in an oscillatory manner). It also makes the airplane less maneuverable on the rolling axis. Lessening dihedral results in better maneuverability on the roll axis, however, the aircraft tends toward spiral instability (continues to roll when upset by turbulence). The rate of divergence is typically low with spiral instability and is easily controlled, however, it can be fatiguing to operate the aircraft in turbulence. Small airplanes designed for maneuverability have no dihedral.

Longitudinal stability may be affected by rigging. A change in incidence of the flying surfaces alters the aircrafts neutral point, the driving factor in longitudinal stability.

There are different aspects of stability and some airplane geometry and rigging is intended to provide or enhance those kinds of stability. See the Pilots Handbook of Aeronautical Knowledge and NACA Technical Note 1670 for a more complete discussion on stability (see the references about NASA and NACA research papers).

Conventional Aircraft in Pitch Equilibrium (Longitudinal Stability)

A conventional aircraft (tail in back) in flight, is being forced to rotate pitch down by the wing and rotate pitch up by the horizontal stabilizer/elevator (Figure 2-13).

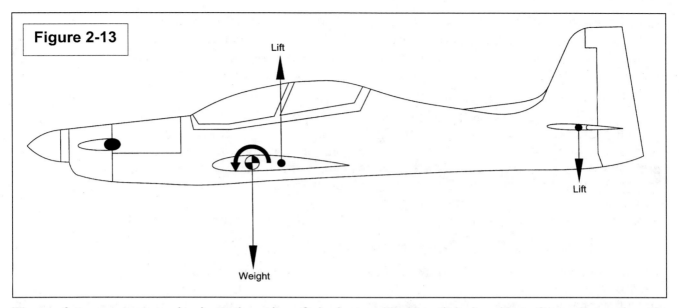

Figure 2-13

These forces are equal when the aircraft is in equilibrium (trim). Aircraft are intentionally designed like this because it provides stability. Changes in speed will change *where* the aerodynamic forces are concentrated, and their *magnitude*, and it will require a change in elevator deflection (or horizontal stabilizer incidence) to compensate. The basic effect of this is that a change in speed will cause a change in pitch of the aircraft, to attempt to maintain the same speed that it was originally [trimmed] at. This is one of the qualities that makes airplanes controllable when disturbed from equilibrium. The downside is that the downward lift produced by the horizontal stabilizer is detracting from the overall performance of the aircraft (its like adding weight) and the wing must be operated at some higher angle of attack to make up for that downward lift. The CG can only be shifted so far back in an attempt to relieve the tail download, because the tail requires a certain amount of leverage (moment) to provide the necessary pitching moments for stability and control (and the CG must be ahead of the lift vector of the wing to provide normal control forces). Some aircraft, in certain phases of flight (flaps down for example), may require a temporary upforce on the horizontal stabilizer to maintain equilibrium.

NOTE

Canard aircraft are considered to be more efficient in this respect because stability is achieved with both the wing and canard producing lift upward.

Slip, Skid, and Sideslip

These terms are given to mean different things when discussing piloting technique, lateral stability, rigging condition, etc.. The terms are used somewhat loosely even in formal textbooks. Since sometimes the same words are used to describe different concepts, they must be taken in context.

Slip and skid, as they occur in turning flight, are an unbalance of centrifugal and lift forces. This is mainly a function of piloting skill on the rudder and a more detailed explanation is given in reference 22. This phenomenon doesn't have much to do with rigging (except in a few airplanes that may have a rudder-aileron interconnect system), it is common to all aircraft.

The term *slip* may be used to describe a number of different conditions, both in flying technique and aircraft behavior. Slip is often used synonymously with *sideslip.*

In discussing piloting techniques, sideslips are used to land in crosswinds and forward slips are used to lose altitude quickly (see the FAA Airplane Flying Handbook). Sideslips and forward slips are about the same in terms of the control inputs, but the magnitudes vary to the situation. In this context, slip is often used as a short form of *sideslip* or *forward slip*.

In flight mechanics and aerodynamics, sideslip (or slip) describes the condition of straight flight when the aircraft is not pointed in the direction it's going, relative to the airflow. The wind axis (X_W) is not in the plane of symmetry of the body axis (X_B) (Figure 2-10 and 2-11). When discussed in terms of lateral stability, the sideslip results in downward acceleration that provides a rolling moment to right the aircraft from the bank, when dihedral effect is present. In rigging, only the stable (unaccelerated) sideslip is considered, the wings are banked in the opposite direction that the aircraft is pointing. This angle of bank is small at slow speeds, and as the speed increases more angle of bank is needed to maintain the same amount of sideslip.

Inclinometer or Slip/Skid Indicator

The inclinometer or slip/skid indicator (Figure 2-14) not only tells of the quality of a turn (balanced forces), but it can also show the magnitude of lateral translation (sideslip) when an aircraft is in straight flight. It is an important instrument in determining the rig of the aircraft.

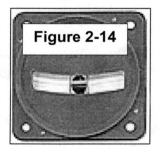

Figure 2-14

An inclinometer is so named because it is used to determine the angle of something that it is attached to, in relation to earth's gravity. It is not used for this purpose in an airplane.

The inclinometer is a level, except that it works in the opposite direction. The ball is heavier than the fluid and so acts in the direction of the greatest forces acting on it (the fluid in the inclinometer is used to damp the ball motion which would otherwise be unreadable because of acceleration, vibration, and turbulence).

During a coordinated turn, the acceleration forces are concentrated along the Z_B+ axis and so the ball is forced to the lowest point of the vial relative to the Z_B axis (ball in the center). In an uncoordinated turn, centrifugal and gravitational forces combine to act on the ball to displace it from the center, in the direction of greatest force. During unaccelerated flight, the ball seeks gravity (Z_E+ axis), the lowest point in the tube, regardless of the attitude of the aircraft.

In the context of rigging, it may be used to give an indication of the alignment of the longitudinal axis of the airplane to the relative wind, *indirectly*. Two common rigging problems are illustrated by the ball. In the first case if the ball isn't centered and the aircraft is in unaccelerated flight, the aircraft must be banked and yawed in the opposite direction to balance the vectors, and therefore any displacement of the ball from center is the result of aircraft attitude (assuming the ball was installed correctly). The second case in rigging is that the ball is slightly out of center with the wings perfectly level, and the aircraft changes direction very slowly (accelerated flight). In this case centrifugal force is forcing the ball to one side. Many slip/skid indicators won't reflect very small changes in forces or attitude (sometimes because they are hard to see). Installation of slip/skid indicators is discussed in Chapter 5.

Coefficients

Coefficients are used frequently in many fields of science and the term is used a lot in the discussion of aircraft science. The term is used in frequently in the discussions that follow so some explanation is provided.

Coefficients are numbers that are established in order to make quantifiable comparisons between different things, those things being dependent on the same variables. It is a relative number, a number without units (dimensionless).

Coefficients are frequently used to describe relative amounts of lift and drag. It is a mathematical method of comparing two aerodynamic bodies which are shaped differently and so perform differently. The concept is most easily demonstrated by an example;

There are two wings with the exactly the same planform and area, but each has a different airfoil. The two wings are put into a wind tunnel at the same angle of attack and wind velocity. Each wing is found to produce a different amount of lift (measured in pounds), indicating that one airfoil is working differently from the other. To compare the two airfoils, an equation is formed using all the variables to come up with a relative number (the coefficient) that can be used to compare the airfoils to each and other airfoils. This number is known as a coefficient, in this case the lift coefficient (denoted C_L). The lift equation for getting the C_L looks like this;

$$C_L = 2 \cdot \frac{L}{V^2 \cdot S \cdot \rho}$$

Where;

L = the measured lift (in pounds)

V = the velocity (feet per second)

S = the wing area (square feet)

p = the air density (slugs per cubic foot)

Drag (C_D) is derived in a similar way;

$$C_D = 2 \cdot \frac{D}{V^2 \cdot S \cdot \rho}$$

Because the lift and drag of an object depend on the speed, air density, shape of the body, size of the body, the angle of the relative wind, etc., coefficients are be used to hold those variables constant and compare only the performance of a shape of a body.

Another way of thinking about a lift or drag coefficient is that it is; the ratio of aerodynamic pressure felt by the component, to the total pressure in the wind.

NOTE

Lift and drag are vectors that are measured in a direction perpendicular and parallel to the relative wind, respectively, and don't really indicate how much *total* force is being reacted by the component. The concept of *total aerodynamic force* is described in more detail later because it is helpful in understanding the interactions between the various parts of the aircraft.

Airfoils/Wing Sections

Airfoil geometry is discussed in detail in Chapter 1.

Symmetrical Airfoil

Symmetrical airfoils are not as efficient at producing lift as asymmetrical airfoils, but they produce lift equally in both directions making them useful for horizontal and vertical stabilizers, and aerobatic aircraft that maneuver in all directions. The chordwise distribution of air pressure around a symmetrical airfoil is illustrated in Figure 2-15.

Note that the position of the lift vectors don't change on symmetrical airfoils with changes in angle of attack. The greatest pressure difference however, is near the leading edge.

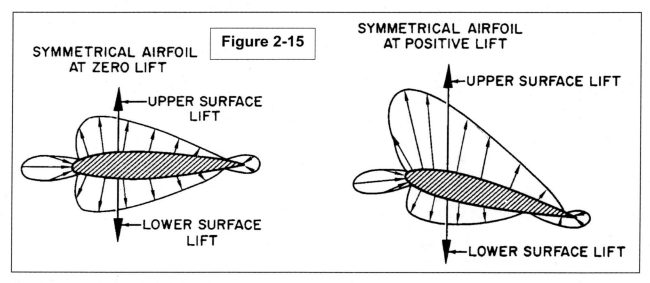

Asymmetrical Airfoil

Asymmetrical airfoils have more camber on one side and produce lift most efficiently in the direction of the most camber. The chordwise distribution of air pressure around a symmetrical airfoil is illustrated in Figure 2-16. Note that the upper and lower force vectors are not in line as they are for the symmetrical airfoil. The forces are attempting to rotate the wing leading edge downwards. The positions of the greatest pressure differences on asymmetrical airfoils depends on the shape of the airfoil.

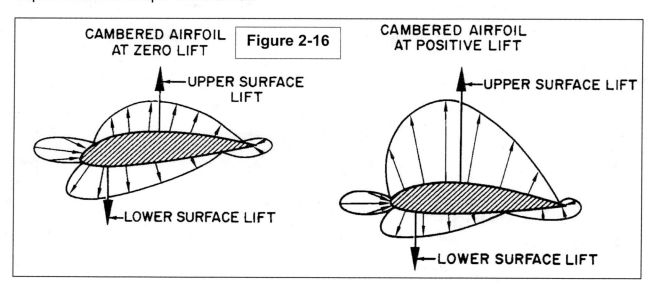

Lift Curve Slope

Five airfoils are depicted in the graph of C_L versus angle of attack in Figure 2-17. Three are symmetrical (the root of the line starts at 0° angle of attack), and the other two are asymmetrical (they still produce lift at 0° angle of attack and to produce no lift requires a negative angle of attack, see about *zero lift line* in Chapter 1). Notice that each airfoil makes about the same angle of line on the graph (until stall). Although the five airfoils depicted here are very different in shape and ability to produce lift, they all change lift by about 0.1 C_L for each 1° angle of attack (the slope of the lift curve). This is the case for almost all the airfoils used on small airplanes.

The aspect ratio of a wing will change the lift curve slope and that is discussed shortly.

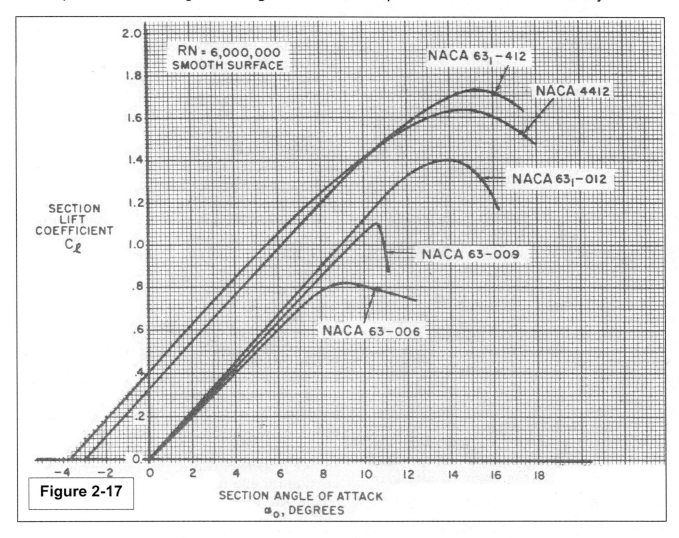

Figure 2-17

Total Aerodynamic Force

Total aerodynamic force is the single reaction of the aircraft to airflow, measured in pounds of force (Figure 2-18). It is a single force vector that illustrates the reaction to the airflow, like direction the hand takes when held out the window of a moving vehicle. Total aerodynamic force is divided into two vector components to evaluate aircraft performance; **lift**, measured perpendicular to the relative wind, and **drag**, measured parallel to the relative wind. Sometimes this manual will only refer to the total aerodynamic force where it is appropriate.

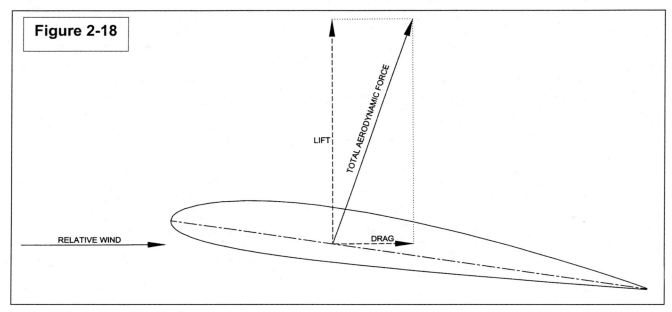

Figure 2-18

The pressure created by an airstream is;

$$q = \frac{1}{2}\rho V^2$$

where;

q = pounds per square foot

ρ = air density, slugs per cubic foot

V = velocity in feet per second

To account for the shape of a body, another coefficient is used, C_F, coefficient of aerodynamic force. So the aerodynamic force produced by a body in an airstream is given as;

F = C_FqS

Where;

F = aerodynamic force

C_F = coefficient of aerodynamic force

q = dynamic pressure

S = surface area of the object seen by the airflow

Notice this formula is the same as the lift and drag formulas given in the paragraph on coefficients, but it's been rearranged a little. This formula doesn't provide the direction of the total aerodynamic force. The direction of the total aerodynamic force depends on the shape of the surface and varies with angle of attack, but can be resolved at any instant if one knows the lift and drag of the part reacting the aerodynamic force.

NOTE

No vectors are actually solved in rigging here, the concept is illustrated as a way of thinking about the effect of changes to the various parts of the aircraft.

As airspeed is increased, the total aerodynamic force increases *as the square of the airspeed*. Put another way, doubling the airspeed will result in four times the aerodynamic force, and to go four times as fast will cause an increase in the total aerodynamic force by a factor of sixteen. Knowing this relationship can be helpful in diagnosing rigging problems as will be seen later.

Lift

Lift is the upward component of total aerodynamic force (Figure 2-18). It is defined as the force perpendicular to the relative wind. Lift is caused by the shape of a wing (the fuselage also produces significant amounts of lift). Lift increases with an increase in angle of attack, or, an increase in airspeed when the angle of attack is held constant. Observe the graphs in Figure 2-19. Lift increases approximately linearly with an increase in angle of attack, however, changes in speed increase the lift very rapidly when the angle of attack is held constant.

Doubling the angle of attack at a constant speed approximately doubles the lift. Doubling the speed at a constant angle of attack increases the lift to four times its value.

For a whole aircraft in unaccelerated flight, angle of attack and airspeed are considered mutually dependent. High speed occurs at low angle of attack and low speed occurs at high angle of attack, the two can't be separated. In rigging it becomes possible to separate angle of attack and airspeed, since one or more lifting surfaces can be adjusted to a different angle than the rest of the aircraft, and will produce greater or lesser reactions with changes in speed. For example, when one lifting surface is off at a different angle from the others, an increase in speed will have a large effect on the aircraft, where as an increase in angle of attack may have very little effect.

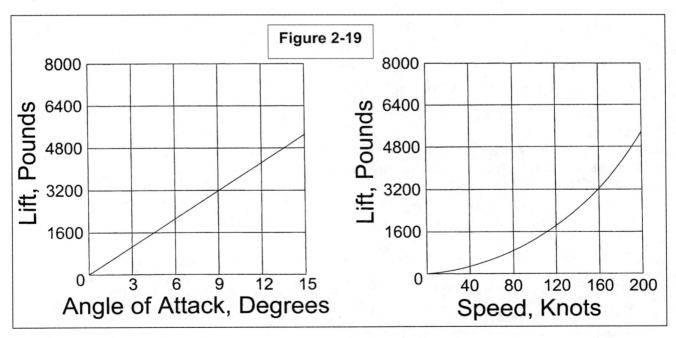

Figure 2-19

Spanwise Lift Distribution

The lift distribution over the span of a wing is not constant. There is more lift at the root than the tip as illustrated in Figure 2-20. For a rectangular wing, the difference in lift between the root and the tip may be almost thirty percent. The elliptically shaped distribution of lift in Figure 2-20 is *approximately* the same for all the typical wing planforms used on small aircraft, however, it does vary with wing planform and aspect ratio. Wash-out (Chapter 1) will also alter this distribution considerably. Rigging changes made inboard on the wing, for example flaps, will have a greater effect on total lift that those made near the wingtips (the rolling moment that is produced by asymmetry here may not be obvious because inboard changes in lift have a smaller moment than outboard changes in lift). To fully understand these relationships for a particular aircraft, the reader is referred to the references at the end of this manual.

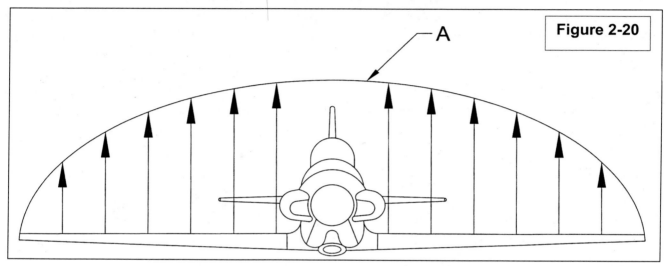

There is also a spanwise movement of air along the wing that increases with angle of attack. The bottom of the wing has the air moving from root to tip with the opposite situation on top of the wing. The spanwise movement is greatest closer to the tips and reduces to zero at the aircraft centerline. This spanwise movement is associated with the wingtip vortices.

Lift Curve Slope of a Whole Wing

Note in Figure 2-17 that the slopes of the lines for the various airfoils are all the same. The airfoils found on most aircraft all have about the same slope; they increase about 0.1 C_L for each one degree of angle of attack. This holds true in two dimensional flow, but when the airfoil is used on a wing (three dimensional flow), the aspect ratio of the wing determines the slope of the line.

NOTE

Two Dimensional and Three Dimensional Flow

Airfoil performance information is provided without regards to the shape of the wing to which it is a part of. The performance of the airfoil will be different when used on a wing on an airplane. Performance of an airfoil is given in the same coefficients (lift and drag) as those used to describe the performance of an actual whole wing, however, the lift coefficients of airfoils are for two-dimensional airflow (referred to as an infinite wing) and the performance will generally be less when it is measured on a wing of a certain shape and size (called a finite wing, or three-dimensional flow). The drag coefficients of airfoils are also for two-dimensional flow and only reflect the form drag of the airfoil (the resistance felt when moving a solid body through a viscous medium). Drag due to lift (induced drag, discussed later) only exists when the airfoil is part of a [finite] wing. In three dimensional flow there is spanwise movement of the air and the wingtip vortices. It is customary in aerodynamic texts to assign lower-case subscripts to two dimensional flow and upper-case subscripts to three dimensional flow, C_L vs. C_l, C_D vs. C_d, etc.. Airplane rigging is mainly concerned with three-dimensional airflow. A number of factors affect the three-dimensional performance of the lifting surface, but aspect ratio has the greatest effect. Planform, taper ratio, and sweepback also affect the characteristics of lift and drag of a wing.

Aspect ratio is the span divided by the average chord, and it defines the three dimensional airflow characteristics about the wing. A glider has a very high aspect ratio (good lift-drag characteristics) while a fighter jet has a very low aspect ratio (poor lift-drag characteristics). In figure 2-21, the effect of aspect ratio on the lift slope is illustrated for five wings with the same airfoil. Aspect ratio is the largest influence on drag due to lift, as will be discussed shortly.

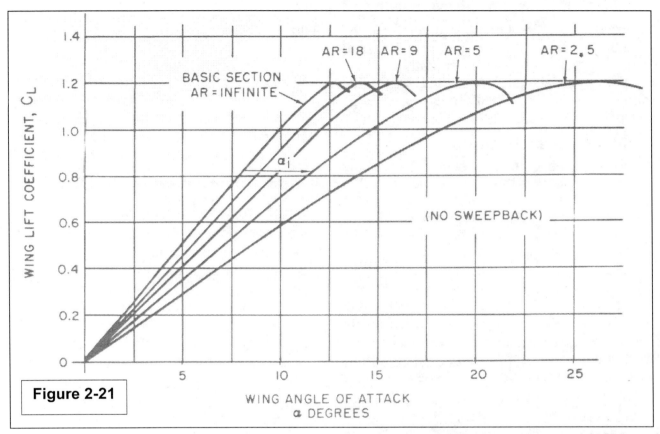

Figure 2-21

Note that as aspect ratio decreases, the surfaces stall at higher angles of attack. This is one way that airplane designers keep the vertical and horizontal stabilizers from stalling.

The actual slope may be estimated from the formula (see Reference 1);

$$\alpha = \frac{0.1R}{R+2}$$

Where;

α = Angle of attack, degrees

R = Aspect ratio

For example, a wing of aspect ratio six will change C_L by .075 for each one degree of angle of attack. This formula doesn't work on very low aspect ratio wings like tail surfaces.

Drag

Drag is the horizontal component of total aerodynamic force. It is defined as the force parallel to the relative wind. Drag is further broken down into classifications that define what is contributing to the total drag force. Just as lift produces aerodynamic forces that affect how an airplane stabilizes in flight, so does drag. From the perspective of rigging, there are two concepts related to drag that are important to rigging.

The first one, trim drag (or rather the reduction of it), has already been discussed briefly. Trim drag is the result of all of the aerodynamic forces required to stabilize the aircraft and doesn't isolate any one force that retards the aircraft or part of the aircraft.

The drag of individual components effects the stabilized attitude of the aircraft. Because there are lifting surfaces hung out all over the place, far from the center of gravity, the drag forces produce moments about the aircraft. Drag as a retarding force is broken down into essentially two parts, parasite and induced drag, which behave slightly differently from each other. An understanding of the difference can help in understanding and troubleshooting rigging problems.

Profile or Parasite Drag

Profile or parasite drag is simply the resistance felt anytime a solid object is placed in an airstream. It follows the square rule of total aerodynamic force in that the resistance increases as the square of the airspeed (doubling speed increases drag by a factor of four and quadrupling speed increases drag by a factor of sixteen). Parasite drag is the dominate drag in high speed flight (low angle of attack). A larger object will produce more drag than a smaller object for a given speed, and a flat object will produce more drag than a streamlined object.

Parasite drag *decreases* as the aircraft slows down. The formula for parasite drag is given in the paragraph on coefficients.

Induced Drag

Induced drag is caused by the production of lift of a wing. On a lifting surface in steady state flight, induced drag *increases* with a *decrease* in airspeed (increase in angle of attack), because of the increased downwash at larger angles of attack and the loss of air at the wing tips (vortices).

For a whole airplane, induced drag changes approximately as the square of the airspeed, but it depends on the airplane design and may increase faster. Many airplanes stall shortly below the minimum drag point and this effect may not be easily recognized during flight, but it is important to recognize that this exists. The graph in Figure 2-22 shows the increase in induced drag of various wings with the same airfoil, as lift is increased.

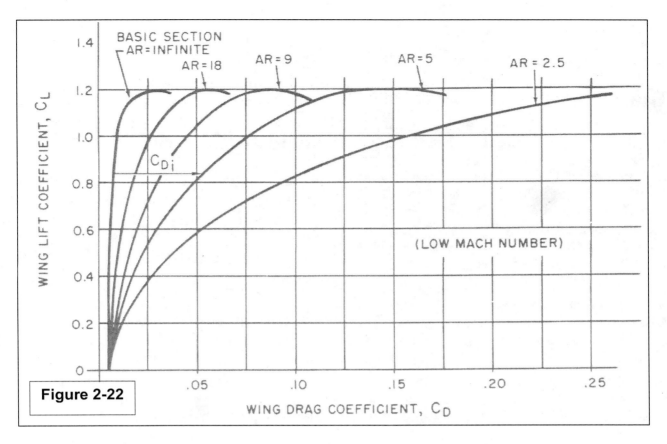

Figure 2-22

Washout

Washout (see Chapter 1) may be used to provide several advantages.

It is used to force the wing root to stall before the wing tip. This will allow some rolling effectiveness and roll stability while the inner part of the wing is stalled, and possibly provide some pilot warning in the form of buffeting on the horizontal stabilizer. It can be used to provide acceptable stall characteristics on a wing which may have had unacceptable characteristics when it was without washout. This is not to mean it is desirable on all airplanes (it may hinder a snap roll). Stalls are discussed in more detail later in this chapter.

Washout may also be employed to provide a more even lift distribution over the span of the wing (see Figure 2-20 for a typical spanwise lift distribution), in doing so reducing the *induced* drag. It can only be set to provide an even lift distribution (minimum induced drag) at one angle of attack though (climb or cruise for example), and will increase drag at other angles of attack.

Geometric washout is adjustable on some airplanes by lengthening or shortening the wing struts/supports, in effect forcibly twisting the wing to a new shape. For most airplanes it is fixed during construction of the wing.

Effect of Aircraft Rotation on Lift and Drag

An aircraft rotating on an axis will cause the relative wind on the lifting surfaces to change direction as the aircraft is rotating. For example, consider an airplane that is rolling. The up going wing will see a decrease in angle of attack, causing a decrease in lift and induced drag. The down going wing will see an increase in the angle of attack, causing an increase in lift and induced drag. This can be critical on control surfaces that are already operating at their maximum ability.

Stall

Stall is the separation of airflow from an airfoil or flying surface because of *excessive angle of attack*. Lift drops off in a stalled condition. The airfoil type (particularly the radius of the leading edge) and wing planform will determine how fast the lift drops off during the stall (notice the drop off for the 63-009 airfoil compared to the other ones in Figure 2-17).

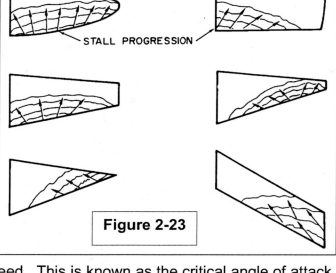

Figure 2-23

A given wing will always stall at the same angle of attack regardless of weight and airspeed. This is known as the critical angle of attack. Most wings on small airplanes stall at approximately 15° - 20° angle of attack, depending on the airfoil and shape of the wing. For a given airfoil, a wing with a low aspect ratio will stall at a higher angle of attack than a wing with a high aspect ratio.

Even though a given wing will normally stall at the same angle of attack all the time, it is possible to delay the stall to a bit higher angle of attack by rotating the wing quickly (pitch up rapidly). The effect is that the air remains attached for a bit longer because of its momentum.

Wing planform and washout also contribute to the reaction of the wing during a stall. The planforms in Figure 2-23 are without washout. Many airplane designs find it desirable to maintain aileron control at very high angles of attack, so some washout is used to keep the stall from immediately spreading to the ailerons.

A wing may also stall when it's close to its' critical angle of attack because of aerodynamic interference from other parts of the aircraft (stall strips do this intentionally, covered later).

It is generally desired that the stall (airflow separation) begin at the wing root and progress to the tip. A wing that stalls at the tip first may roll violently without any warning. The rapid roll rate keeps one wing unstalled and the other wing completely stalled, aggravating the situation. Large yawing moments are also produced that are pro-spin. The tip stall tends to occur without warning because the separated airflow does not cause buffeting at the tail. Stall strips (Chapter 4) are used to force the wing root to stall before the tip, and provide buffeting to warn the pilot of the impending stall.

It is also desired that the progression through the critical angle of attack (hence airflow separation) be somewhat gradual, requiring several degrees change in angle of attack to make the airflow become completely separated. Certain airfoils (those with a small leading radius) may begin to stall at the leading edge rather than the trailing edge, causing a very abrupt loss of lift at the stall.

It is also undesirable that a stall progress along the trailing edge towards the tips, before connecting to the leading edge. This may result in a complete and sudden loss of lift. Stall strips help with forcing stalls to occur at the wing root, rather than progress towards the wing tips along the trailing edge.

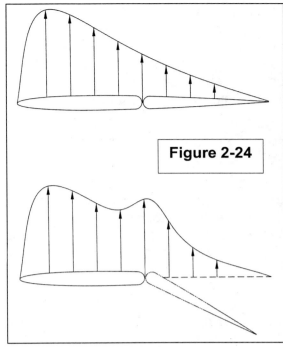

Figure 2-24

Some tailoring of the stall characteristics of the wing are possible to ensure that the stall begins at the wing root, but most of them are an aspect of aircraft design. (using some washout in the wing or a different airfoil at the tips). The use of washout results in a decrease of the maximum lift that a wing is capable of producing. It is also possible that a stall occurring at the wing root will interfere with the airflow over the tail, to make the aircraft difficult to control (see deep stall phenomenon in this chapter).

In aircraft used for aerobatics, gradual stall progression from the wing root outwards may not be desired as it hinders rapid maneuvers between normal flight and stalled/partially stalled flight.

Stall fences are employed on some aircraft to delay spanwise progression of the separated airflow. Fences have also been found useful at one or both ends of the ailerons to increase control authority and control power over a wide range of angles of attack. Sizing of such fences is out of the scope of this book. They may exert a powerful influence over the flying qualities of small aircraft, both positive and negative.

Deflection of Flaps on Wings

Ailerons, elevators, and rudders are frequently termed as *flaps* when discussing the aerodynamics of lifting surfaces that are so equipped.

The chordwise distribution of lift changes with flap deflection (Figure 2-24). The upper half of this figure shows the same pressure distributions in Figure 2-16, but the vectors were removed from the bottom surface and added to the top.

Lift increases approximately proportionally (linearly) with flap deflection, within the normal range of deflections. Ailerons, elevators, and rudders generally do not deflect more than 30° because of airflow separation, non-linear responses to control deflections, and rapidly diminishing returns of lift versus drag.

Both parasite and induced drag are increased when a flapped surface is deflected. Induced drag because of the increase in lift due to camber, and parasite drag due to the increase in area shown to the relative wind.

The angle of attack at which a wing will stall is *reduced* by the deflection of flaps (remember that ailerons, elevators, and rudders are also considered flaps in this context). A fully deflected flap (or aileron, rudder, elevator) will reduce the critical angle of attack of the wing to which it is attached by several degrees (Figure 2-25). A wing which normally stalls at fifteen degrees may instead stall at twelve degrees with the control surface fully deflected (that span of the wing that has the surface attached to it). This is important in setting the maximum control deflections or troubleshooting problems. It can be aggravated by the fact that when an airplane is rotating about an axis, the change in direction of the relative wind on that surface(s) can produce a stall, or delay it.

Stalling of the controls may produce different reactions. There may be a lightening or reversal of control pressure as airflow separation occurs. The rotation rate of the aircraft may suddenly decrease or stop altogether. Buffeting may be felt through the control actuator. It may be difficult to stop a rotation with opposite control pressure because the aircraft rotation is increasing the angle of attack of that surface. If the horizontal stabilizer stalls, the aircraft may rapidly pitch down or tumble (discussed again later in this chapter and Appendix B).

Sometimes airflow separation can be induced by other factors (maneuvering, turbulence, aerodynamic interference) where the surface was previously on the verge of separation but not quite there.

Stall of the vertical stabilizer is indirectly related to allowable rudder travel. It may occur at large angles of yaw when the destabilizing force of the deflected rudder (or propeller) is greater than the stabilizing force of the vertical stabilizer. The restoring forces of the vertical stabilizer/rudder combination do not increase beyond about 15° of sideslip, however the yawing moments of the propeller and fuselage that oppose stability continue to increase to about 45° of sideslip.

NOTE

Even when the vertical stabilizer is not stalled, a reversal of rudder force (becomes easier to deflect as it is deflected further) may occur at low speed and high power at some sideslip angle, due to lack of total directional stability (a design problem). Pilots refer to this is as a hard-over rudder.

For many airplanes there is no yaw without roll (inducing some yaw angle will cause a roll in the same direction) and the behavior of an airplane undergoing stalling of the vertical stabilizer will be different when roll is present. The size of the fixed portion of the vertical stabilizer (including dorsal/ventral fins) in relation to the rudder plays a major role in whether stalling occurs, but excessive deflection angles may allow this to occur as well. Reducing the allowable deflection of the rudder may cause problems elsewhere.

The deflection of flaps changes the position of the zero lift line of the airfoil (see Airfoil Geometry in Chapter 1 about the zero lift line). An airfoil which produced zero lift at an angle of attack of -3° may have a zero lift angle of attack of -10° or more when the landing flaps are fully deployed. Symmetrical airfoils have a zero lift line equal to the chord line, however that zero lift line also changes with flap deflection.

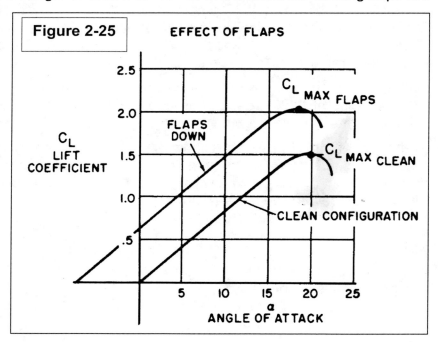

It is common for control surfaces to have different positive and negative deflections. Rudders usually deflect symmetrically (except for some propeller airplanes with large engines) but ailerons and elevators may deflect asymmetrically, for different reasons.

Many aileron systems have different deflections to try and counteract the adverse yaw phenomenon. Adverse yaw is the tendency of an aircraft to yaw in the opposite direction of bank. The yawing moment increases proportionally with the lift coefficient being developed by the wing (gets worse with increasing angle of attack). It occurs for several reasons;

- The induced drag of a wing with the aileron deflected downwards becomes greater, and the induced drag of the wing with the up aileron becomes less.

- The lift vector of the down going wing becomes tilted slightly forward, while the up-going wing's lift vector becomes tilted slightly aft. This is because lift is perpendicular to the relative wind. The relative wind arrives from a different direction when the aircraft is rolling.

- The lift and drag forces are concentrated more outboard on the wing with the down aileron (and the opposite for the up aileron), creating a larger yawing moment.

To help balance the forces which create adverse yaw, many ailerons systems will have different deflections between up and down. This requires a differential linkage system to make one aileron go down differently than the other one going up.

Elevators typically deflect more in the up direction than the down direction. More authority is needed in the up direction, to trim the airplane at a slow speed and for takeoff and landing with an aft CG and with flaps deployed. Down elevator deflection is required to trim an airplane for high speed, or in the case of some airplanes, to compensate for flap deployment. In a broader sense, the human organism doesn't tolerate negative accelerations very well (down elevator), so airplanes are designed to mainly operate in the envelope of positive load factors (up elevator). Stabilators generally deflect far less than elevators to achieve the same result.

The neutral position of ailerons which are long or full span will affect the aerodynamic characteristics of the wing. A slight droop of the ailerons may add lift but also drag and a pitching moment which will require elevator to trim out. The ailerons and flaps on some aircraft are designed to be reflexed upward in flight with a cockpit adjustment, to reduce the drag and trim requirements in high speed flight.

More of a design factor, but easily altered, is the shape of the trailing edge of the control surface. Much change (both positive and negative) can be made to the behavior of a control by altering the shape of its trailing edge. Much research is available on this subject (free NASA and NACA research papers, see References). In addition, convex or concave shapes on the control surface, intentional or otherwise, can have a large effect on the behavior of the controls.

Effect of Center of Gravity on Aircraft Performance

The longitudinal location of the center of gravity can strongly influence the aircrafts performance, even when it is held within the limits of the design. When evaluating performance changes due to changes in rigging, it is necessary to consider the CG location in two ways;

1) When evaluating changes in aircraft performance due to rigging changes, comparisons must be made at the same weight, weight distribution, and CG.

2) When evaluating aircraft controllability after rigging changes, it must be evaluated at the extreme limits of the weight and CG range.

Much of the following text is taken directly from the Pilots' Handbook of Aeronautical Knowledge.

Performance Aspects of CG

The effect of the position of the center of gravity may be very significant to climb and cruising performance. Refer to Figure 2-26.

With forward loading, "nose-up" trim is required in most airplanes to maintain level cruising flight. Nose-up trim involves setting the tail surfaces to produce a greater down load on the aft portion of the fuselage, which adds to the wing loading and the total lift required from the wing. This results in a higher stalling speed. The added weight requires a higher angle of attack of the wing, which results in more drag and a lower cruising speed.

With aft loading and "nose-down" trim, the tail surfaces will exert less down load, relieving the wing of that much wing loading and lift required to maintain altitude. The reduced weight results in a lower stall speed. The required angle of attack of the wing is less, so the drag is less, allowing for a faster cruise speed.

Theoretically, a neutral load on the tail surfaces would produce the most efficient overall performance and fastest cruise speed, but would also result in instability. Consequently, airplanes are designed to require a down load on the tail for stability and controllability.

Controllability and Stability Aspects of CG

The effects of the distribution of the airplane's useful load have a significant influence on its flight characteristics, even when the load is within the center-of-gravity limits and the maximum permissible gross weight.

Generally, an airplane becomes less controllable, especially at slow flight speeds, as the center of gravity is moved further aft. An airplane which cleanly recovers from a prolonged spin with the center of gravity at one position may fail completely to respond to normal recovery attempts when the center of gravity is moved aft by 1 or 2 inches. It is common practice for airplane designers to establish an aft center-of-gravity limit that is within 1 inch of the maximum which will allow normal recovery from a one-turn spin. When certificating an airplane in the utility category to permit intentional spins, the aft center-of-gravity limit is usually established at a point several inches forward of that which is permissible for certification in the normal category.

Another factor affecting controllability is the effect of long moment arms to the positions of heavy equipment and cargo. The same airplane may be loaded to maximum gross weight within its center-of-gravity limits by concentrating fuel, passengers, and cargo near the design center of gravity; or by dispersing fuel and cargo loads in wingtip tanks and cargo bins forward and aft of the cabin. With the same total weight and center of gravity, maneuvering the airplane or maintaining level flight in turbulent air will require the application of greater control forces when the load is dispersed. This is

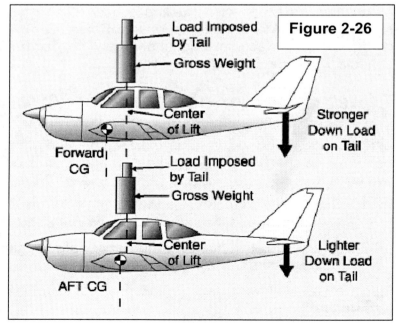

Figure 2-26

true because of the longer moment arms to the positions of the heavy fuel and cargo loads which must be overcome by the action of the control surfaces (Ed. note, the rotational moments due to inertia initially may make an aircraft appear more stable because of its resistance to change directions, however, once it starts to rotate it is more difficult to arrest the rotation). An airplane with full outboard wing tanks or tip tanks tends to be sluggish in roll when control situations are marginal, while one with full nose and aft cargo bins tends to be less responsive to the elevator controls.

The rearward center-of-gravity limit of an airplane is determined largely by considerations of stability. The original airworthiness requirements for a type certificate specify that an airplane in flight at a certain speed will dampen out vertical displacement of the nose within a certain number of oscillations. An airplane loaded too far rearward may not do this; instead when the nose is momentarily pulled up, it may alternately climb and dive becoming steeper with each oscillation. This instability is not only uncomfortable to occupants, but it could even become dangerous by making the airplane unmanageable under certain conditions.

The recovery from a stall in any airplane becomes progressively more difficult as its center of gravity moves aft. This is particularly important in spin recovery, as there is a point in rearward loading of any airplane at which a "flat" spin will develop. A flat spin is one in which centrifugal force, acting through a center of gravity located well to the rear, will pull the tail of the airplane out away from the axis of the spin, making it impossible to get the nose down and recover.

An airplane loaded to the rear limit of its permissible center-of-gravity range will handle differently in turns and stall maneuvers and have different landing characteristics than when it is loaded near the forward limit.

The forward center-of-gravity limit is determined by a number of considerations. It should be possible to trim the aircraft at its' slowest flight speed with the power off. A tailwheel airplane must be capable of a full stall, power-off landing. A tailwheel airplane loaded excessively nose heavy will be difficult to taxi, particularly in high winds. It can be nosed over easily by use of the brakes, and it will be difficult to land without bouncing since it tends to pitch down on the wheels as it is slowed down and flared for landing. It may be difficult to rotate a nosewheel type aircraft for takeoff, or flare for landing, with a CG that is too far forward. Steering difficulties on the ground may occur in nosewheel-type airplanes, particularly during the landing roll and takeoff.

Higher elevator control forces normally exist with a forward CG location due to the increased stabilizer deflection required to balance the airplane.

The airplane becomes less stable as the CG is moved rearward. This is because when the CG is moved rearward it causes an increase in the angle of attack. Therefore, the wing contribution to the airplane's stability is now decreased, while the tail contribution is still stabilizing. When the point is reached that the wing and tail contributions balance, then neutral stability exists. Any CG movement further aft will result in an unstable airplane.

A CG located more forward results in greater stability (more leverage available from the horizontal stabilizer), but increases the need for greater back elevator pressure. The elevator may no longer be able to oppose any increase in nose-down pitching. Adequate elevator control is needed to control the airplane throughout the airspeed range down to the stall.

Effect of Propeller on Airplane Dynamics

The propeller creates considerable unfavorable forces that need to be trimmed out to keep the airplane flying in a desirable manner. Many airplanes are rigged to help make some compensation for these forces. In discussing the forces produced by the propeller, the concept of propeller *disk* is often used to describe the rotational plane of the propeller, illustrated in Figure 2-27.

Undesirable propeller forces typically become worse with higher thrust levels and/or an increased angle of attack of the aircraft. All of the propeller effects will vary with the speed of the aircraft because at some high power setting, greater thrust exists at lower speeds (remember an aircraft accelerates until thrust equals drag and an aircraft in equilibrium has no excess thrust until power is increased). The deflection of wing flaps also tends to magnify the effects.

Yawing Moments and Sideslip

The most obvious propeller forces to the pilot are those that yaw the aircraft and are discussed first. Yawing moments are generated by the rotational velocity imparted to the slipstream by the propeller (spiraling slipstream), and, asymmetrical thrust (P-factor). For an airplane with a propeller that spins clockwise as viewed from the rear, the moments cause a left yaw.

Spiraling Slipstream

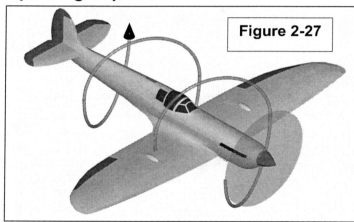

Figure 2-27

The magnitude and effect of the spiraling slipstream varies greatly between aircraft designs, as such, no easy rules of thumb can be provided to isolate these forces from other propeller phenomenon. It is the result of the air circulating around the aircraft because the propeller imparts such motion to it (Figure 2-27).

The rotating air changes the direction of the local airflow at the side of the fuselage and vertical stabilizer, causing a yawing moment to the left (for clockwise prop rotation), illustrated in Figure 2-28 (it may also contribute to a pitching moment and rolling moment, discussed later). This phenomenon varies with aircraft design, power setting, angle of attack, and airspeed, both in effect and magnitude.

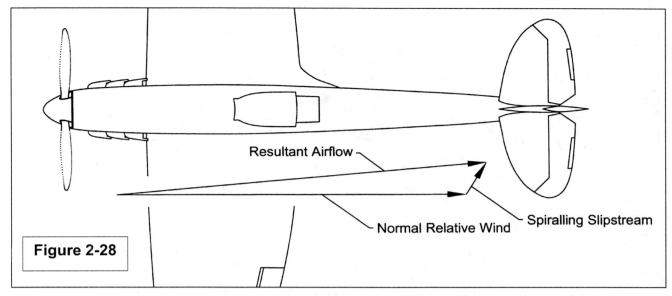

Figure 2-28

The yawing moments produced by the propeller are mainly caused by the spiralling slipstream. It depends on the airplane, but for conventional configurations about half of that yaw is caused by the slipstream striking the fuselage aft of the CG, and the other half is from the slipstream striking the vertical stabilizer. Wind tunnel research on conventional single-engine aircraft indicate that about 6/7 of the total yawing moment is produced by the spiraling slipstream and the rest is from asymmetrical thrust, discussed in the next section.

Much rudder is often needed to compensate in some flight conditions, so much so that very little is available for maneuvering (less control authority). Although spiraling slipstream occurs constantly, it imparts more influence to the airflow at slow speed and higher power settings. On most certificated aircraft, it's been partly rigged out (discussed shortly). The magnitude of this rigging can sometimes be seen by diving the aircraft as fast as it can go safely, and it will probably require significant *left* rudder to keep the ball centered at high speed.

Compensation for spiraling slipstream doesn't eliminate the problem, it is a compromise between efficiency in a desired flight condition and sufficient control authority.

Asymmetrical Thrust (P-Factor)

When the airflow into the propeller isn't perpendicular to the propeller plane or disc, the thrust produced isn't symmetrical about the disc. One half of the disc produces more thrust and the other half produces less. This asymmetrical thrust is also known as P-Factor. For an aircraft in straight flight but pitched up slightly, the down going blade pushes more air back than the up going blade. The reason for this is illustrated in Figure 2-29. The propeller on the bottom is viewed directly from above the aircraft when the aircraft has its longitudinal axis pointed into the wind. The propeller on the top is also viewed from above the aircraft, but the airplane is pitched up in relation to the relative wind. Note the size of the blade on the right side of the prop (down going blade) compared to the size of the blade on the left side (up going blade).

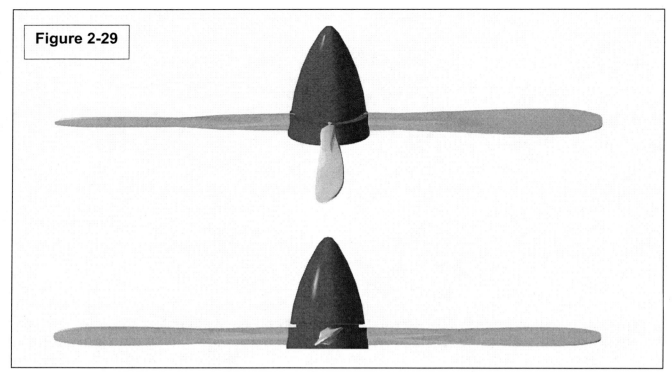

Figure 2-29

The reaction to this is a left yaw of the aircraft (for a prop which rotates clockwise as viewed from the rear) because the thrust being produced on the right side of the aircraft (down going blade) is about twice as much as the up going blade on the left side (Figure 2-30) (the converse effect is that if an aircraft is flying in slightly yawed flight, a pitching moment is produced).

The magnitude of this effect is greater with increasing horsepower, propeller size, and pitch. Nothing really can be done about this because it will occur anytime the propeller is producing thrust and the relative wind is at an angle to the propeller disc. P-factor also produces a vibration, because the loads on the propeller blades are constantly fluctuating. Three and four bladed propellers reduce the overall vibration due to P-factor.

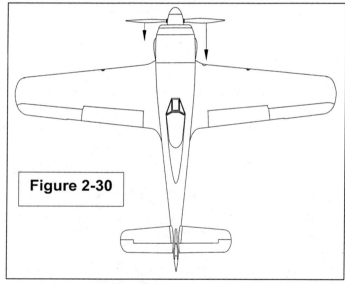

Figure 2-30

More Yawing Moments

Torque and gyroscopic precession also cause aircraft reactions.

Torque is a relatively light force in a small plane with an average engine, but does help create differences in roll rate from one direction to the other. Because aileron deflection may be required to overcome torque, the asymmetrical aileron drag results in yaw and requires more rudder deflection.

Gyroscopic precession will only affect an airplane while it is *changing* attitude, and this can be hard to separate from asymmetrical thrust which also starts producing a yawing or pitching moment as soon as the aircraft starts to rotate in pitch or yaw.

A thrust axis which isn't aligned with the longitudinal axis of the aircraft may also contribute to a yawing moment or pitching moment, or both.

The Overall Effect of Yawing Moments

Right rudder is needed to counteract the left yaw produced by spiraling slipstream and asymmetrical thrust, but then the situation gets more complicated. When right rudder is used to prevent yaw, it results in left translation of the aircraft (relative to the earth) because the vertical stabilizer/rudder is now producing a left side force (left side of Figure 2-31). To prevent left translation of the aircraft, some more right rudder must be used to offset the left side force (right side of Figure 2-31). The aircraft is now going in the correct direction, but is translating with respect to the airflow. In addition, if the airplane were banked slightly to the right to offset torque, more right rudder will be required (from adverse yaw).

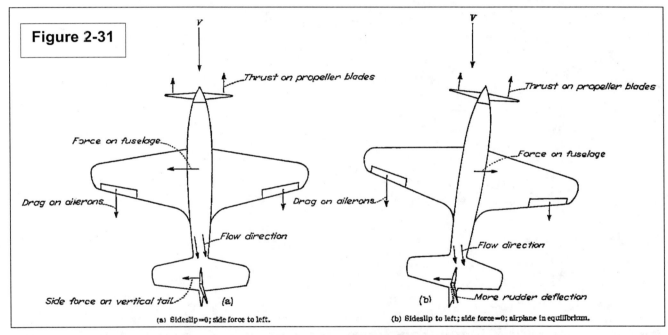

(a) Sideslip = 0; side force to left.

(b) Sideslip to left; side force = 0; airplane in equilibrium.

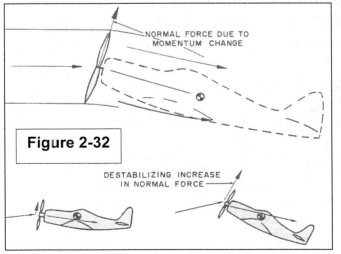

Pitching Moments

The propeller and its thrust cause a number of pitching moments. They may occur during a power change, speed change, and in steady flight, especially at high power and high angles of attack. Deflection of wing flaps tend to magnify these effects. Pitching moments from the propeller aren't always obvious to the pilot because airplanes require frequent elevator adjustments anyway due to speed changes, configuration changes, etc..

Moment of Propeller Normal Force About the Center of Gravity

Propellers produce an upward lift component (perpendicular to the thrust line) due to the altering of airflow as it passes through the propeller (Figure 2-32). Because of this, a pitching moment is produced that requires some trim to counteract during flight. For a conventional tractor airplane, it attempts to rotate the aircraft nose up and gets worse as the aircraft angle of attack and power are increased. This is a relatively small force compared to the other propeller forces that attempt to pitch the aircraft.

Moment of Propeller Axial Force About the Center of Gravity

When the line of action of the thrust (propeller axial force) passes above or below the vertical CG of an aircraft, aircraft pitching moments are produced when thrust exists (Figure 2-33). The magnitude of this effect increases with angle of attack and thrust. It may be most noticeable during large power changes.

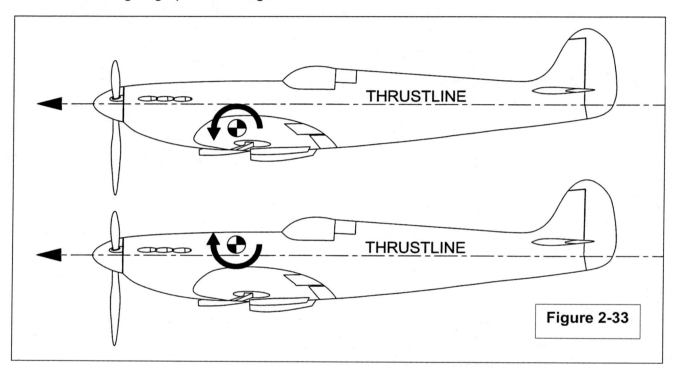

Figure 2-33

Increased Angle of Downwash, Increased Dynamic Pressure at the Tail, and Change in Pitching Moment of the Wing Due to the Action of the Slipstream

These effects are all related and interact with each other to produce various moments on the aircraft.

An increase in propeller normal force (discussed earlier) increases the downwash angle of the wing relative to the incoming airflow, thereby changing the angle that it strikes the horizontal stabilizer. The change in angle of attack of the stabilizer changes its lift and creates a pitching moment.

The increased velocity of the air at the horizontal stabilizer increases its effectiveness.

The increased velocity of the air at the wing changes its pitching moment, requiring more elevator deflection.

The most common overall effect of these three phenomenon is to produce a nose-up moment of the aircraft at slower speeds and higher thrust levels. These effects tend to reduce the longitudinal stability of the aircraft.

On some aircraft configurations, the action of the spiraling slipstream may strike the horizontal stabilizer and cause a pitching moment, similar to the yawing moments discussed earlier. For most airplanes though, the influence of the wing downwash on the horizontal stabilizer is so great that any action of the spiraling slipstream on the horizontal stabilizer is lost in the airflow and the effect is negligible. A sideslip may also cause a pitching moment, which is usually nose down in tractor airplanes.

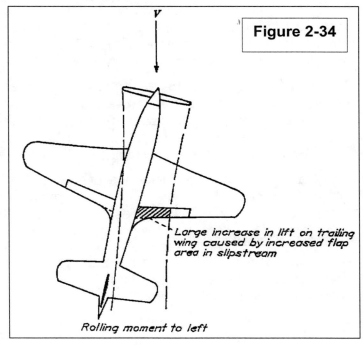

Figure 2-34

Tilting the thrust axis so that it points downwards one or two degrees (such that the thrust tries to rotate the aircraft nose down) has shown to make a significant increase in the stability of the aircraft when it is at high power and low speeds, however, this results in a certain amount of trim drag at high speeds.

Rolling Moments

A sideslip may induce a rolling moment with power-on because the slipstream strikes more wing (Figure 2-34). Engine torque may contribute to a rolling moment depending on the size of the engine/prop in relation to the aircraft. Spiralling slipstream may contribute to a rolling moment if it alters the direction of the airflow over the wing.

Rigging for Propeller Phenomenon

This is one of the biggest hassles of rigging since no matter how well everything is done, airplanes with propellers want to follow a different path. Some things may be done to minimize the effects of the propeller during certain flight conditions, while other things may be done to enhance performance. If the powerplant is really big in relation to the rest of the airplane, some things will have to be done just to maintain some semblance of control.

Experimenting with the thrust axis to create yawing or pitching moments may significantly improve performance or handling in a particular flight regime. Proper rigging of the thrustline can reduce trim drag and increase control authority in a relatively narrow band of the aircraft flight envelope, or change stability in another. Because of the number of factors involved, it is impossible to provide guidelines as to the proper angles, and some requirements for control are directly in contrast with the desire for reduced trim drag.

On average, a propeller loses about 2% efficiency when the inflow of the air to the propeller disk is not perpendicular to the disk, for angles less than ten degrees, and the loss in efficiency increases rapidly for greater angles. The overall loss in propulsive efficiency is small compared to the pitching or yawing moments that are produced, which require trim (drag) to cancel out.

Yaw

The most pronounced effect of the propeller phenomena is yaw. The rudder must be powerful enough to provide equilibrium of yawing moments at any flight condition or speed. If the aircraft is a compromise between maneuvering and cruise, it is desirable to reduce the trim drag as much as possible but retain sufficient control authority to coordinate a maximum rate roll to the right (for a clockwise propeller) with full power, at minimum controllable airspeed. Because rudder is the primary roll control when the aircraft is stalled, it should be possible to prevent post-stall roll (incipient spin), with the power off. An aerobatic airplane generally needs more rudder authority than other airplanes.

Two methods are available to provide the necessary control authority in yaw during high-power/high angle of attack maneuvering; offset vertical stabilizer and offset thrust. Wind tunnel tests have shown that increasing the area of the vertical stabilizer generally makes the yawing effects worse (making the rudder bigger may cause a hard-over rudder, vertical stabilizer stall, or reversal of control force, discussed earlier).

Any change in either the incidence of the vertical stabilizer or the thrustline will be good for only one phase of flight, creating moments which need to be trimmed out at other speeds. Some compensation may be necessary though, to provide sufficient control authority.

Vertical Stabilizer

Setting the vertical stabilizer at an angle to the longitudinal axis will realign it with the relative wind and allow more authority from the rudder, since the rudder needs not be deflected (or deflected as much) to coordinate flight (Figure 2-35). Drag is also reduced for this particular flight condition.

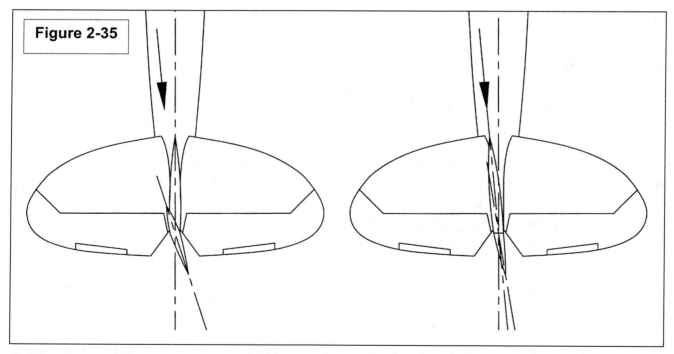

Setting the stabilizer to compensate for the low-speed/high power flight condition is going to cause a yaw to occur in the high-speed flight condition, increasing trim drag and changing the directional stability. Refer to Figure 2-36 to see the results of wind tunnel tests done on a 1940's era fighter. Although the speeds are considerably higher, the effects are similar.

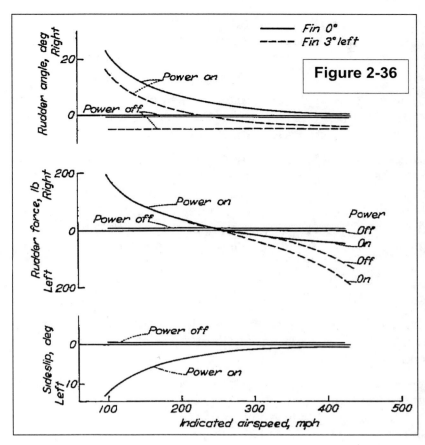

Figure 2-36

An additional problem on faster aircraft caused by setting the vertical stabilizer at an angle is the possibility of distortion of the vertical stabilizer/rudder at the higher speeds, especially on fabric covered aircraft.

Although the angle of the vertical stabilizer is fixed during construction, some adjustment may be possible by reworking the airframe. One possibility, used on the successful Messerschmitt ME109 (a fast WWII propeller driven fighter), is an unsymmetrical airfoil for the vertical stabilizer/rudder.

Thrustline

Setting the thrustline at an angle to the centerline of the aircraft (such that it tries to pull the aircraft to the right for a clockwise propeller) produces both a yawing moment and changes the effect of the slipstream. Because of this, small changes to the thrustline may have very large effects in countering the yaw. However, an aircraft with a fairly large speed envelope may suffer greatly in performance, in addition to increasing vibration from P-factor. The thrustline is generally easier to change than the incidence angle of the vertical stabilizer.

Offset CG

NACA research indicates that a 2% offset of lateral CG may reduce required rudder deflection by as much as ten degrees on a high-powered airplane.

Pitch

The thrustline is generally oriented in pitch to keep the prop efficient in its intended flight regime (relative wind approximately perpendicular to the prop disk). This sometimes should be compromised with the pitching moments that are produced (see the previous paragraphs). A thrust axis inclined downward will reduce the load on the horizontal stabilizer (trim drag) and increase stability in high-power/low-speed conditions. It may be worthwhile experimenting with the angular orientation of the thrustline (viewed from the side) to see if some performance gains are possible. When changing the orientation of the thrustline it is wise to do it in 1° increments, testing in between modifications for performance and stability changes. Moving the thrust line upward (nose-up pitching moment) is generally more critical than downward because of the decrease in low speed static stability.

Roll

Pure rolling moments generated by the propeller slipstream are usually fairly small and there is generally sufficient aileron authority to overcome them with plenty of reserve. Some configurations may suffer more than others, and because adverse yaw accompanies aileron input, rudder authority is lost and trim drag increases. There is little that can be done to compensate for low speed rolling moments without causing large rolling moments in high speed flight.

There is a large difference in roll rate in a propeller driven airplane, between making rolls to the left or right, partly due to engine torque. This can't be avoided, if the prop were big enough the airplane would need an anti-torque propeller on one wing, similar to a helicopter.

One noticeable effect of the propeller on the roll axis is during the power-on stall. It may be pronounced enough to cause the airplane to always roll in one direction. This may be caused by an asymmetry of high-velocity air due to the slipstream, or it may be caused by a slight slip occurring during the stall as in Figure 2-34. This can sometimes be corrected with asymmetrical placement of stall strips (see Chapter 4), however that may cause asymmetrical stalls to occur with power off. In any case, experimentation may produce a satisfactory compromise.

Incidence Angle of the Wing

The driving factor in determining the incidence angle of the wing is the intended use of the aircraft (see incidence angle in Chapter 1). The incidence angle of the wing is closely related to the incidence angle of the horizontal stabilizer.

Consider the following; for some speed that the aircraft is operating at, there is only one angle of attack that will support the aircraft (at some given weight). If the wing were set at that angle to the fuselage (incidence angle), the fuselage would be pointed into the relative wind at that airspeed. This will produce the minimum amount of drag. Note this only works at one airspeed (and weight). In Figure 2-37, it has been determined that the wing on this 1000 LB airplane will produce 1000 LBS of lift at the design airspeed of 200 KTS, at a 2° angle of attack. So the wing is set with an incidence angle of 2° to the longitudinal axis of the fuselage, to provide the minimum frontal area of the fuselage to the relative wind.

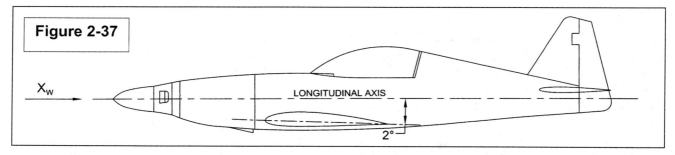

To complicate things, it has been shown in wind tunnel testing of fuselages that the minimum drag of the fuselage does not always occur when the longitudinal axis is aligned with the relative wind. Some fuselages make minimum drag when they are pitched down a degree or so, some just the opposite. While it is fairly easy to calculate the incidence angle for a wing, the fuselage is something else. If the incidence angle of the wing is adjustable, experimenting with this aspect of rigging may be well worth it in terms of performance gains for a particular aspect of flight. For a racer, one would pick the maximum expected airspeed to minimize the parasite drag, for a cross-country plane one might use the expected cruise speed to reduce fuel consumption, an aerobatic plane may use maneuvering speed to increase control authority and minimize the loss of kinetic energy from drag while maneuvering, a STOL airplane will be optimized for takeoff and landing etc..

Note that the geometric definition of incidence angle is the difference between the chord line of the wing and the longitudinal axis of the aircraft. In determining the optimum incidence angle, the difference is really between the zero lift line of the wing and the longitudinal axis of the aircraft. The zero lift line of the airfoil is often a few degrees different from the chord line (see Chapter 1), and the zero-lift line of a whole wing may be different than that of the airfoil. It is beyond the scope of this book to calculate all of the factors, but simple techniques are provided in Reference 17. Since only a wind tunnel or sophisticated CFD program is going to predict the minimum drag fuselage angle, experimentation is going to produce the best results.

Changing the incidence angle of the wing will affect the attitude that the airplane stalls at, which is important to taking off and landing because the gear geometry holds the allowable ground attitudes within certain limits. It is desired that the airplane will produce lift at a reasonable rotation angle for takeoff, to keep the takeoff speed fairly low. It is also desirable to be able to attain a reasonable pitch angle at a fairly slow speed since a normal landing should be performed close to, or at the stall speed. This can be particularly important in taildraggers. If the taildragger must stall at a very steep pitch attitude for landing, the tailwheel strikes first, creating control problems. Some tailwheel aircraft are designed so that the pitch angle in the three-point attitude is about two degrees less than the stalling angle of attack to *prevent* a full-stall landing. This is generally the case on aircraft which have poor stall characteristics and tend to roll rapidly during a stall.

Aircraft which cannot exceed the critical angle of attack on the ground are termed *geometry limited*, the tail will hit the ground before the critical angle of attack is reached. In the example in Figure 2-38, assume the aircraft has a stalling angle of attack of 14°. The airplane in the upper part of the picture will strike the tail on the ground before the critical angle of attack, therefore it will not be possible to over-rotate it on takeoff without hitting the tail. The airplane in the lower part of the picture has a taller gear, and it is possible to rotate the aircraft on takeoff to a stalling angle of attack (never lifting off except for possibly the influence of ground effect).

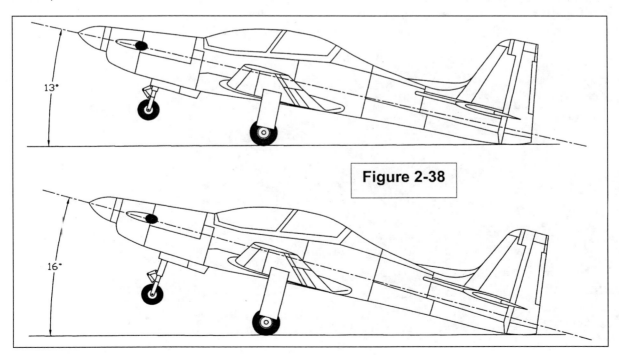

Figure 2-38

Aircraft are made both ways, and either case may be acceptable if the landing speeds are appropriate.

The angle that the aircraft makes on the ground (and therefore the angle that the wing makes on the ground) affect the takeoff and landing speed, and therefore the takeoff/landing distance. The effect of small changes in takeoff and landing speeds are nonlinear in terms of the distances required.

Consider the following relationship of takeoff/landing speed versus distance required;

$$\frac{S_2}{S_1} = \left(\frac{V_2}{V_1}\right)^2$$

Where;

S_1 = Takeoff or landing distance corresponding to some original speed

S_2 = Takeoff or landing distance corresponding to some new speed

V_1 = Takeoff or landing speed corresponding to some original distance

V_2 = Takeoff or landing speed corresponding to some new distance

Try working the numbers and it will be seen that a ten percent change in velocity is going to change the required distance by more than twenty percent.

STOL aircraft are most sensitive to the gear geometry relationship. The ability to leave the ground initially plays a large part in the total takeoff distance, and to minimize the ground roll on landing requires a touchdown at the stall speed. NASA experimented with a blown-wing STOL transport (Figure 2-39) that could jump off the ground by charging the nose strut suddenly at takeoff speed, from an onboard accumulator. This helped eliminate some of the required elevator power required to rotate the aircraft on takeoff because on the ground, the aircraft rotates about the main wheels and not the CG. When the main wheels are behind the CG (nosegear aircraft), the elevator has a shorter arm (less leverage).

Figure 2-39

Just because a particular airplane design was well tested doesn't mean the optimum incidence angle was determined. If the prototype airplane was controllable and produced satisfactory performance, a designer may not fool around with it, particularly if it's difficult to change.

Changes in wing incidence angle must be accompanied by changes in incidence of the horizontal stabilizer (or neutral position of stabilators). The relationship between the two lifting surfaces can be critical from a control and stability perspective and is discussed next. The downwash from the wing that impinges on the horizontal stabilizer is a significant force. A change in the downwash angle may create a critical situation regarding the angle of attack on the horizontal stabilizer in some phase of flight, resulting in a loss of elevator effectiveness or a complete stall of the horizontal stabilizer.

The thrustline may also be changed to accommodate a new fuselage angle into the relative wind. Because the propeller forces and the lift forces from the fuselage create pitching moments, a thrustline parallel to the relative wind and/or longitudinal axis of the aircraft may not be the optimum solution.

Incidence Angle of the Horizontal Stabilizer

This section applies to conventional aircraft only (tail in back). Canards work differently, but some of the principles are similar. Rigging subtleties of canards are out of the scope of this book.

The incidence angle of the horizontal stabilizer is determined by a number of factors, which must be compromised to each other to provide aircraft control throughout the intended envelope. This is not predicted as easily as the incidence angle of the wing. In many aircraft the adjustment of the stabilizer incidence is not available after construction, and therefore changes in incidence angle of the wing (if possible) will be limited and must be carefully tested.

As the wing produces lift, the air leaving the trailing edge is deflected downwards. The amount of downwards deflection of the air increases with an increase in angle of attack and flap deflection. The downward movement of the air behind the wing affects the angle of attack of the horizontal stabilizer. Because of this effect, any change in angle of attack of the wing is followed by a lesser change in angle of attack at the horizontal stabilizer (e.g. a 10° change in wing angle of attack may result in a 6° change at the horizontal stabilizer). If the horizontal stabilizer is mounted higher than the wing, this effect may be reduced.

Some aircraft use the whole horizontal stabilizer for trim (Chapter 1). Other aircraft control pitch with a stabilator (Chapter 1). Most airplanes have a horizontal stabilizer that is fixed in position and the amount of pitching moment that the elevator can generate is very finite, especially in the wake of large wing flaps. Small changes in incidence (1°) can have large effects on aircraft control. The deflection of the elevator changes the position of the zero lift line of the horizontal stabilizer. It may be possible to reduce the trim drag and increase the control authority for the aircrafts intended flight envelope by experimenting with stabilizer incidence settings, such that the least amount of elevator deflection is required to trim the aircraft at the intended flight regime. The elevator deflection required for trim varies approximately as the square of the airspeed (*Stable* graph in Figure 2-40).

Changes in wing incidence will normally be accompanied by a change in stabilizer incidence and consequently the deflection limits of the elevator (if the control stops are on the fuselage rather than the horizontal stabilizer). One can play it safe by simply keeping the relationship the same from what was known to work. For example if the difference between the wing and stabilizer incidence was +2°, then an adjustment of the wing incidence would require an adjustment in stabilizer incidence to maintain the +2° relationship. This is a simple approximation and won't hold for much change because the change in the angle that the fuselage is hitting the relative wind will alter its pitching moment, causing a total change in the trim requirement of the aircraft. For aircraft with all movable horizontal stabilizers, whether for trim or pitch, their neutral position and deflection limits should change to accompany a new wing incidence angle.

Where no rigging information is given for the incidence of the horizontal stabilizer, it is *usually* safe to *initially* set the stabilizer incidence at 0° (chord line parallel to the longitudinal axis of the airplane). That assumes a stabilizer with a symmetrical airfoil, as most are. The aircraft should be carefully flight tested to ensure that it meets certain requirements, given in the next section.

The asymmetrical airfoil used for the horizontal stabilizer and elevator on some aircraft is particularly sensitive to changes in incidence. One must be particularly cautious in changing the incidence angle of an asymmetrical section from what is known to work.

Some STOL aircraft (or aircraft equipped with STOL kits) begin to lose elevator effectiveness before the airflow on the wing begins to separate, limiting the approach speed/angle of the aircraft because of loss in control authority, particularly in turbulence. Some benefits may be had then by experimenting with the incidence(s) in that particular phase of flight, however, it will be a compromise between other phases of flight and the aircraft must be carefully flight tested with regards to the test points given in the next section.

The longitudinal stability of an aircraft is affected by the incidence angle of the horizontal stabilizer. It is beyond the scope of this book to get into the details of airplane stability, but the stability tests given in FAA Advisory Circular <u>AC 90-89, Amateur-Built Flight Testing Handbook</u> should be carried out.

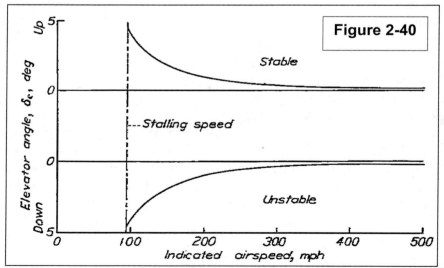

Figure 2-40

One design factor related to the incidence of the horizontal stabilizer is that it must not stall in the flight envelope of the airplane. Airplanes are designed such that even when the wing is stalled well beyond its critical angle of attack, sufficient pitch control authority is available. In normal flight, the horizontal stabilizer/elevator produces lift downward (discussed earlier in the chapter). If the horizontal stabilizer/elevator is highly loaded (operating at a high angle of attack), any factor which increases its angle of attack may cause it to stall if it is not adjusted properly. A number of design factors influence the ability of the horizontal

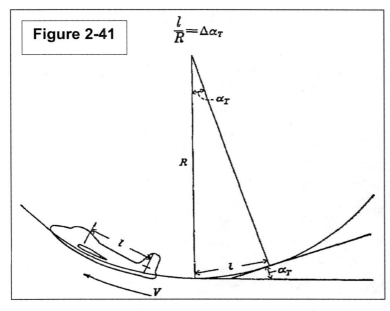

Figure 2-41

stabilizer to remain unstalled, but a change in incidence of either the wing or horizontal stabilizer can significantly alter the ability of the horizontal stabilizer to continue producing lift under some conditions. If changes are made to the incidence relationship in order to increase the performance of the airplane, experimentation should be conducted in small increments with flight testing done at each change. Testing for this is discussed later in this chapter.

Canard aircraft are designed just the opposite of conventional airplanes, the canard is designed to stall before the wing.

Effect of Incidence Changes on Maneuvering

Changes in incidence angle of the wing or horizontal stabilizer will affect the ability to maneuver the aircraft in accelerated (turning) flight. The curvature of the flight path in accelerated flight reduces the angle of attack at the horizontal stabilizer, requiring more elevator deflection to continue accelerating at a constant rate (Figure 2-41). An incidence angle that is far off will not allow the aircraft to develop its maximum lift, or may require more elevator power than is aerodynamically possible.

When comparing the differences between incidence changes, it is necessary to evaluate the turning ability of the aircraft in level or descending flight. An aircraft which simply pulls up from level flight is already experiencing a 1g acceleration which must be added to the acceleration of the pull-up. Therefore a straight pull-up at a constant g produces a curve with a greater radius than if the aircraft is rolled into a 90° bank, or upside down, and turned at the same g loading. The straight pull-up pictured in Figure 2-41 then results in a lesser change of angle of attack at the horizontal stabilizer and doesn't reflect the whole flight envelope of the aircraft in question.

Minimum Flying Qualities to Be Investigated After Incidence Changes of the Wing or Tail

Incidence changes have a powerful effect on flying qualities and performance and some recommended requirements are given here. While some of these may seem arbitrary, there are far-reaching consequences of not having sufficient control of the airplane.

It is desired that an airplane return to the trimmed condition in one cycle of motion when the elevator is deflected and released.

It should be possible to develop either the maximum load factor or maximum lift coefficient at any flying speed.

It should be possible to control pitch in both minimum speed and maximum speed flight, at forward and aft CG positions, flaps up and down, power full and at idle.

For takeoff there should be sufficient pitch authority to hold the aircraft in any attitude while still on the ground at 80% of takeoff speed for nosewheel aircraft, or 50% of takeoff speed for tailwheel aircraft. The takeoff and landing is often the most critical with regards to elevator control because ground effect alters the downwash angle at the tail and makes the plane more stable (requires more up elevator to trim). Flap deployment may also greatly affect the downwash over the horizontal stabilizer.

The elevator should have enough control power to hold a tailwheel airplane off the ground in a three point attitude. For nosewheel airplanes, sufficient elevator power must exist to hold the aircraft off the ground in the landing attitude (approximately the same as a level power-off stall). There should be some elevator power over and above the previous requirements to provide control authority in turbulence. The previous checks should be performed at forward and aft CG positions, with takeoff or landing flaps as appropriate.

These requirements make quite a large matrix of test points, but it is important to ensure the aircraft will remain in control in expected service conditions.

Stall of the Horizontal Stabilizer

It should not be possible to stall the horizontal stabilizer during any normal flight condition. A wing stall and stabilizer stall may at first appear to be the same when experienced from the cockpit, however, there are differences. In a wing stall, the aircraft pitches down *relatively* gradually, resulting in a curved forward flightpath. When the horizontal stabilizer stalls, the aircraft *immediately* rotates pitch down about its' CG when the horizontal stabilizer downforce is suddenly lost. The aircraft ends up pointed in a completely different direction than it's going. Although it resembles a wing stall, the recovery procedure is exactly the opposite; pull back on the stick to reduce the angle of attack at the horizontal stabilizer.

Incidence is only one factor in the ability of the horizontal stabilizer to resist stall (aspect ratio, airfoil, and configuration of the horizontal stabilizer are some of the design factors).

Any factor which increases the tail download is more likely to make the horizontal stabilizer stall; flap extension, forward CG, higher speed, and maneuvering (pushover at less than 1 g). For a fixed horizontal stabilizer and movable elevator, the speed at which the stall will occur is higher than with a horizontal stabilizer that changes incidence for trim. The effects of power are airplane specific; increasing engine power on high thrust line aircraft (thrust line above the CG) requires more nose-up trim and is more likely to cause stall. In addition, the propeller slipstream increases the strength and angle of wing downwash, especially with flaps extended. Sideslip may also induce the stall on some configurations of empennage (T-tail and cruciform).

Flight testing for this should done at an altitude not less than one would practice wing stalls. It would be wise to wear a parachute. Because it might be possible to get a particular airplane turned around in relation to the relative wind with the right combination and timing of elevator input, it is suggested to initiate recovery (aft stick) at the first sign of stall. Most empennages aren't structurally capable of taking the airflow from the rear.

Signs of impending stall are;

- difficulty to trim in the pitch axis.
- a pulsing or buffeting of the longitudinal control.
- a lightening of control push force (or an increase in pull force) necessary to command a new pitch attitude.

On some airplanes. the airflow around the elevator during the stall can push the elevator trailing-edge down with such great force that it exceeds the pilots ability to move the control. Retracting flaps, adjusting power, and/or waiting for the airplane to stabilize in the new attitude may help alleviate the control forces on the elevator.

Be aggressive when trying to make the stabilizer stall; Do zero g pushovers throughout the speed range in various flap configurations and extreme CG. Establish maximum magnitude slips in both directions, power-off and power-on. Oscillate the airplane in yaw with increasing amplitude to the maximum attainable, when the horizontal stabilizer is highly loaded. Gusts and turbulence can drive the horizontal stabilizer to stall when it is already close. It is necessary to be aggressive in finding the stall so that it doesn't come as a surprise later on when close to the ground.

General recovery procedures are; aft stick, flaps up, power to idle.

Bungee and Spring Centering Systems for the Elevator Control Circuit

Some aircraft use bungees or springs in the elevator control circuit to help overcome adverse control loads or to provide more appropriate control forces. Because of the variety of applications, no guidelines are provided here. The correct type of spring (*spring rate*), in addition to the load that it imparts is important as it affects the basic controllability of the aircraft in certain situations. Any spring in the control system can affect the dynamic stability of the aircraft.

Force trim systems, commonly used on rudders, also use a bungee or spring to provide the desired force.

Biplane Aerodynamics

The airflow through and around the wings of a biplane cellule cause one wing to affect the other, and they both affect the airflow over the horizontal stabilizer. The combined wing (the cellule) pitching forces determine the stability and center of gravity range of the biplane.

It is difficult to separate the effects of stagger, wing planform, aspect ratio, etc, in a single chapter, due to the wide variety of biplane configurations and combinations of wing planforms (the upper and lower wings are often different in planform and airfoil, in addition to being rigged with different geometry). NACA has done extensive research on biplane configurations and the effects of wing geometry. That material is available free from the NASA Technical Report Server (see References). Reference 1 contains an introductory chapter on biplane aerodynamics and flight mechanics.

Deep Stall Phenomenon

Deep stall generally only occurs on conventional T-tail aircraft or canard aircraft, for different reasons. On conventional T-tail aircraft, the wake from the stalled wing envelopes the horizontal stabilizer/stabilator with a corresponding loss of effectiveness of the horizontal stabilizer and/or control power of the elevator. Canard aircraft may get into deep stall when the main wing stalls before the canard.

A wing stall is only part of the deep stall phenomenon. The aerodynamic force vectors balance the aircraft in a new angle of attack range (very steep, 45° for example) for which aircraft are not designed to be controlled. It results in the aircraft stabilizing in a very steep descent, but with the aircraft in a relatively level attitude. Sometimes the descent is slow, and in small airplanes the occupants have survived the impact.

The ability of T-tail aircraft to avoid deep stall may be affected by inappropriate rigging of the incidence angles of the wing and horizontal stabilizer, or allowable deflections of the elevator or flaps. Canard aircraft are particularly sensitive to the incidence angle of the canard to ensure it stalls before the wing. The canards of aircraft with conventional controls are of a high aspect ratio to help ensure they stall before the wing. Rigging subtleties of canards are out of the scope of this book.

Figure 2-42
NASA investigating deep stall phenomenon using a glider with a modified horizontal stabilizer.

Chapter 3
Rigging Tools

An aircraft can be rigged with a minimum amount of equipment although a few things may need to be fabricated by the builder. It is typical that a particular model of aircraft requires some special tools for rigging, but they are generally simple things that are used in conjunction with standard tools, and are easily fabricated. There is usually more than one way to rig some aspect of the airplane geometry and some of the tools described here are intended to increase accuracy and/or are easier to use than alternate methods. Some of the alternate methods are discussed in Chapter 5, Initial Rigging.

Metrology for Rigging

Metrology is the science of measuring. Rigging an aircraft involves taking measurements. Exceptional results can be achieved with commonly available measuring tools by understanding their limitations and applying some principles of metrology. There is a point of diminishing returns when trying to achieve a lot of precision since the fabrication tolerances of the airframe may be fairly large. However, the careful application of measuring tools in the presence of airframe asymmetry will result in a faster compromise than guessing, and a lesser aerodynamic asymmetry. In any case, consistency in measuring is important and the more care made in making measurements will result in less frustration when multiple changes to the rigging are expected.

The following discussion not only applies to the instruments one uses, but also to the methods or system.

Accuracy, Resolution, and Repeatability

It's important to differentiate between accuracy, resolution, and repeatability. These are the three standards by which measuring tools and measuring methods are judged, and only the best instruments and systems may possess all three qualities.

Accuracy is that an instrument or system provides meaningful information against some common standard, be it inches or degrees. The tolerances for airplane rigging can be easily met with commonly available inexpensive tools.

Resolution is how fine measurements may be judged. An instrument may be calibrated in .01" increments, or it may be as fine as .0001" increments. Note that an instrument may have fine resolution, but suffers in accuracy. Many cheap tools have very fine divisions (high resolution), but are only useful for rough measurements because of the quality (accuracy).

Repeatability is the ability of an instrument to provide consistent measurements time after time. This is one of the most important properties of a measuring tool or system. An instrument may suffer in accuracy but be consistent in its inaccuracy (always .01" off, for example). This isn't too bad of a situation in many cases as there are ways of compensating if one recognizes where the inaccuracy is and by how much. Even the best instruments may suffer in accuracy at some point of their range, but they are always off by the same amount at the same place, making them predictable.

Airplane rigging requires more repeatability than accuracy for several reasons. One reason is that to make an airplane symmetrical about its centerline, measurements must be made on both sides of the aircraft. It is desired to have everything the same. Another reason is that when experimenting with the rigging to increase performance, it is necessary to make new adjustments based on the previous rigging condition.

When choosing measuring tools and taking measurements, these properties must be considered. Since one can compensate for certain inaccuracies (sometimes) by the Symmetric Distribution of Error (discussed next), one may need only an instrument or system that is repeatable. A good example of a repeatable *system* is one that always uses the same reference datum. The type of job dictates the required accuracy and resolution.

Symmetric Distribution of Error

An important principle of metrology is called the Symmetric Distribution of Error. More simply, divide the difference. This can be applied to all types of measurements. For example, when measuring to find the exact center of something, measure from two opposite sides and make the center mark between the two measurements. Another example is when using a square to make something perpendicular to something else, try putting the square on opposite sides and divide any discernible error. It is possible to achieve very precise measurements in this manner with just average measuring tools. Because aerodynamic reactions are often not proportional to the geometric deviations, the application of this principle tempered by an understanding of the aerodynamics will result in the best solution

Length of Straightedges

Straightedges (including levels) provide the best measurements when their length approximately equals or exceeds the piece that is being measured. This may seem overly obvious but its worth discussing briefly. When making measurements with the level or straightedge it is best to assume the surface of the thing being measured has a number of imperfections, high spots and low spots. A level or straightedge of sufficient length will tend to average out the surface imperfections and provide a meaningful measurement. A level or straightedge which is too small to span the work may only indicate the difference between one high spot and one low spot, giving a false measurement. A tool of sufficient size may not provide an accurate measurement in the presence of large surface imperfections, however, the state of the surface can quickly be inspected visually by looking for light showing between the tool and the surface. It will be obvious if the tool is not mainly in contact with the surface (a human eye can see a gap of less than 0.001" when a light is shown on the back side). Thus it is important to not only use a tool of sufficient length, but to ensure that the instrument is actually averaging out the surface imperfections by looking at the contact areas.

Knowing the importance of length, the principle can be used to put two tools together to make a larger, more useful tool. One good example of this is the setting of a 12" precision level onto a straightedge of longer length, to produce a highly sensitive and long level (see the next section on levels about this). A protractor may also be placed on a longer straightedge in order to span a large control surface, for the rigging of control throws. There are limits to this system since a small tool is generally designed for a small job and will not produce the desired results if used with too large of a straightedge. A good rule of thumb for rigging is that the accuracy and repeatability of the system will be suspect if the small tool is used on a straightedge more than four times the length of the working surface of the small tool. That assumes that the tools were of a decent quality and good working surfaces to begin with.

The same principle applies to squares used in rigging. It is necessary on some installations to place one leg of a square onto a straightedge which is spanning a long surface.

Many measurements on an aircraft are angular. Small angular errors may result in large linear errors, as the distance increases from the root of the measurement (Figure 3-1). It is important to make every effort to span as much surface as possible.

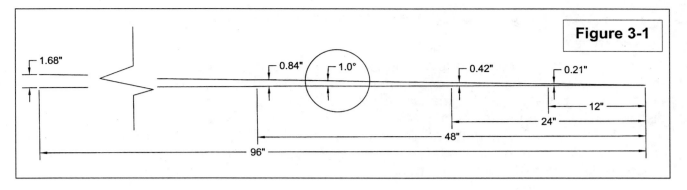

Figure 3-1

Measuring Instruments

Levels

Levels are used extensively in airplane rigging, frequently with some sort of adapter to allow them to be used on curved surfaces. Adapters and aircraft specific rigging tools are discussed later.

Simple levels used in general construction are suitable for rigging. Several sizes (lengths) may be necessary. Certain construction levels have highly curved vials so that the bubble makes less movement under rough conditions of use. This property reduces accuracy, resolution, and repeatability. While it is good enough for concrete forms, it is undesirable for accurate rigging.

Most box beam levels have machined surfaces and tend to be more accurate, straighter, and stronger. The I-beam levels are usually just extrusions (Figure 3-2) and are easily twisted out of shape, even from sitting in the sun. Levels are typically more accurate when in the horizontal position, as opposed to vertical.

Accuracy

Inexpensive levels typically do not have a stated accuracy. Certain brand-name levels that are commonly available do state the accuracy on the level. Those levels are generally held to tighter tolerances during manufacture, particularly with regards to the straightness.

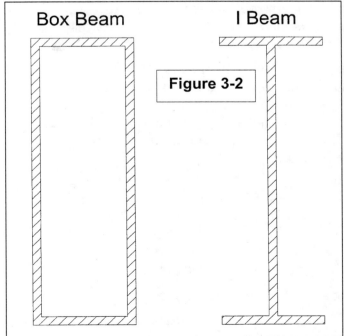

Figure 3-2

A level bubble may be checked for accuracy by placing it on a surface and noting the position of the bubble. If the level is now swapped end-for-end and replaced in the exact same position on the surface, the bubble should be in the exact same position it was originally. It is easier to see a difference if the surface is not quite level, such that the bubble is touching one of the lines initially. Levels are typically calibrated at manufacture such that the vial is up, so the lower edge on the surface to be measured would produce the greatest accuracy.

It seems that all of the common brand name levels that specify accuracy list it as .029°. This is the equivalent of an error of .024" at

48". This is suitable for all rigging jobs on the small aircraft.

Resolution

The resolution of the common level sometimes leaves something to be desired. It really is a matter of how well one can perceive movements of the bubble in relation to the lines on the bubble vial. If the bubble is small in relation to the space between the lines, it is difficult to discern small changes in the movement of the level. The size of the bubble may vary somewhat between levels of the same make and model due to manufacturing tolerances of the vials. Sometimes an extra mark(s) made with a fine point marker on the vial may be helpful for those levels or measurements that need it.

Looking at the resolution versus the accuracy of the finer levels, it will be seen that if a .024" shim is placed under one end of a forty-eight inch level (moves the level by the stated accuracy of .029°), the bubble moves but it is difficult to discern unless the edge of a bubble is near a line. It is discernable if one looks *very* hard. This implies the accuracy is equal to or greater than the resolution, a desirable condition.

Repeatability

A level is repeatable to the limit of its accuracy and resolution. If it can be swapped end for end with no change in the bubble, it is accurate and will give the same reading no matter which way it is placed. Repeatability is more of a user issue than anything. Prior to making precise measurements, wipe the surface of the level and where it's going to be placed before putting it down. When it is to be moved between measurements on the same spot, replace it *exactly* in the position it was previously. When using it to obtain bilateral symmetry, like when setting wing incidence, use the same position(s) on either wing. Due to surface imperfections on individual wings, it may be necessary to take a number of measurements with the level on each wing and average them out, but still placing the level in mirror image positions on the wings. The ability to discern fine changes in the position of the bubble (resolution) plays a direct part in the instruments repeatability.

Accuracy, Resolution, and their Overall Effect on Rigging

This all begs the question then of how much is going to make a difference in rigging. Looked at in terms of angular degrees of resolution, rather than inches, it is difficult to discern a change of .029° movement of a quality four foot level, but still possible. An example of a rigging problem will help illustrate the effect of measuring errors on the flight characteristics of an aircraft.

A typical tolerance for rigging the wing incidence of a small aircraft is ±0.25° between either side. Obviously this is well within the capabilities of the level and that much bubble movement will be easy to discern (it is about ten times less than the accuracy and about ten times more than the resolution). This tolerance however is only an initial rigging tolerance to ensure that the aircraft will remain controllable during the first test flight for rigging condition, and does not represent the optimum situation (wings exactly the same).

To illustrate the effects of small errors an example is given. There is no need to compute all of this for rigging, it is only given to quantify the effect of errors in measurement The example will use an airplane with;

- a perfectly rectangular wing planform
- two ten foot long wings (twenty foot span) with a three foot wing chord
- wings are individually adjustable for incidence and have no washout

Let it be assumed that the wings have been adjusted such that they are both approximately 0.1° in error in the opposite directions, making a total difference of 0.2° between the two wings, well within the rigging tolerance.

Now the difference in lift of the two wings will be estimated at various speeds. The following formula gives the lift of the wing;

$$L = C_L \bullet \frac{V^2}{295} \bullet S$$

L = Lift, pounds

C_L = Coeffcient of lift which relates directly to the angle of attack, explained in Chapter 2 (in this case it is the difference in incidence angle of the wings)

V = Velocity of the aircraft in Knots

S = Area of the wing, each individual wing having 10' X 3' = 30 square feet

NOTE

This is the same lift equation given in Chapter 2, it has only been rearranged to use Knots instead of feet per second, and the air density is taken to be sea level.

Assume as in Chapter 2 that 1.0° of angle of attack is the equivalent of 0.077 C_L for this wing (it has an aspect ratio of 6.7), therefore a 0.1° change in angle is 0.0077 C_L. Putting the numbers in the equation for 100 KTS, each wing produces approximately 8 LBS of lift for each 0.1° of angle of attack.

$$L = .0077 \bullet \frac{100^2}{295} \bullet 30 = 7.8 \text{ LBS}$$

With the wings erred in opposite directions, there is a total of 16 LBS of force difference between the two sides. That 8 LBS of lift is distributed along the length of each wing (not evenly, see Chapter 2) so no rolling or yawing force may be felt (because of control friction perhaps). It does detract from the performance of the aircraft since that lift isn't contributing to anything, and its' reducing control authority in one direction. Look at the case of 150 KTS because the aerodynamic reaction increases with airspeed faster than it does with angle of attack;

$$L = .0077 \bullet \frac{150^2}{295} \bullet 30 = 17.6 \text{ LBS}$$

The force more than doubled with only a fifty percent increase in airspeed, creating an 36 LBS difference in lift on opposite sides of the aircraft. This may cause an objectionable rolling force in a small aircraft. The overall effect on drag and control authority is highly cumulative because each control must be deflected some amount to compensate and hold the airplane in equilibrium.

It is seen that an error of 0.1° for each wing may have been acceptable in the rigging tolerances, however it may have produced objectionable trim requirements and also decreased performance. Plugging the numbers in again using a **total** error between the two wings of .029° (the accuracy of the level if the resolution is sufficient), results in a C_L of .0029 and;

$$L = .0029 \bullet \frac{150^2}{295} \bullet 36 = 8 \text{ LBS}$$

This will probably not produce objectionable flying qualities in a typical aircraft attached to this particular wing planform. It still saps performance though and the techniques given in subsequent chapters on adjusting the incidence help to eliminate the smaller errors. Looked at another way, this difference in lift between the two wings is not contributing to aircraft performance, and equalizing the lift between the two is equivalent to lightening the airplane by the same number of pounds.

The example given above may ask the question, "How is it possible to get the wings within .029° if the accuracy of the level is .029°?. It is possible that each wing is .029° off in opposite directions creating a total error of .058°". The answer is in the application of the level. The accuracy of a level is given in its ability to provide a meaningful measurement regardless of which way the ends are pointing. By always placing the level in the same direction (mark one end and point it forward for example), it will be about twice as repeatable (has half the potential error). In this manner the ability to take measurements is only limited by the resolution. Although the level may still produce an error of .029°, all the measurements taken will be consistent to each other. More likely, surface imperfections in the wing will preclude this amount of accuracy but this inherent inaccuracy can be minimized by taking a number of measurements and applying the symmetric distribution of error..

The rigging tolerances given in maintenance documentation are there to ensure that an aircraft has sufficient control authority. They do not necessarily provide optimum flying characteristics or performance.

Machinist or Precision Level

To achieve the ultimate in accuracy and resolution, a machinist level (Figure 3-3) may be used, usually laid on a longer straightedge to span the necessary parts. Although surface imperfections tend to reduce the overall accuracy of this system, it is still more accurate and most importantly provides a high degree of resolution and repeatability if used correctly. Commonly available machinist levels are 12" or shorter in length, and the bubble glass is indexed with .0010" or .0005" inch graduations. Note that the bubble may go to full scale deflection if one end of the level is raised 0.005", so it is very sensitive.

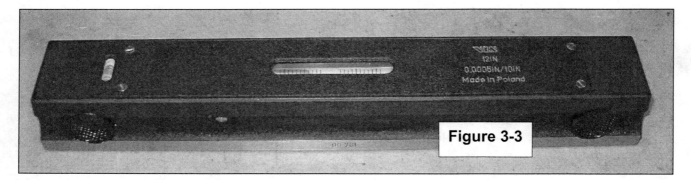

Figure 3-3

Electronic Levels

There are many electronic levels available that display the angular difference from level and perform some other useful functions. Many of them seem to have a higher resolution than their accuracy, giving the impression to the user that they are sufficiently accurate. Also the sensitivity is frequently damped by the software code to enable it to be more useful in a general construction setting (like the levels with the highly curved vials). That has a direct effect on repeatability. Study the specifications and do the tolerance study given above before using it for rigging critical aspects of the airplane. Having said all that, the short electronic levels can be very handy for rigging the control throws where much less of the above properties are required (see Protractors and Combination Heads).

Protractors/Clinometers; Propeller Protractor, Protractor Head, or Angle Finder

There are many common tools designed to provide an angular measurement. In rigging they are used to determine control surface deflection. A protractor measures angles in relation to some other surface, while a clinometer measures angles in relation to gravity (like a level indicates perpendicular to gravity). The terms are often used interchangeably and are herein referred to as protractors.

Since the working surfaces of the protractors are fairly small, they are most suited for short chords like control surfaces. Some surfaces need to be spanned with a straightedge on which the clinometer is placed, in order to get meaningful results. A whole wing or wing chord is most accurately measured with a regular level and incidence/dihedral board to produce the correct angle.

NOTE

Rigging the maximum angular deflection of the primary controls generally requires far less precision than some other aspects of rigging, because only the maximum deflection is being measured. The tolerances for the maximum control surface deflections are fairly liberal because it is difficult to discern the difference in control authority when the controls are fully deflected while one degree off the allowed deflection. There can be some critical issues though on some aircraft caused by excessive deflection like separation of airflow on the control surface in some flight regimes and aerodynamic locking. These are discussed in more detail in Chapter 2.

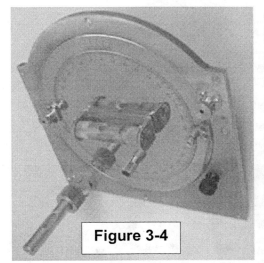

Figure 3-4

A common tool found in the professional aircraft mechanics hanger is the Propeller Protractor (Figure 3-4). They are commonly used for setting control deflections despite the name. The propeller protractor was designed for setting propeller blade angles which require more precision than the control surface throws of most aircraft. They cost about a thousand dollars. They are very accurate and provide a resolution of six minutes. They are only discussed here because they may be encountered in rigging in some shops. They are not necessary for rigging the control surfaces on the average aircraft.

The protractor head from a combination square like the one in Figure 3-5 is well suited for rigging the control deflections. A high quality one will provide a resolution of about one-half degree on the angular dial with comparable accuracy. This tool is commonly found in the workshop but the higher quality ones generally aren't available in stores and must be ordered.

Figure 3-5

A common angle finder like the shown in Figure 3-6 is frequently purchased in hardware stores but should be checked for accuracy before use. Some of them provide a resolution of less than one degree and are not very repeatable because the friction in the pointer bearings causes the pointer to stick.

Electronic angle finders should be evaluated for accuracy, resolution, and repeatability. A good quality one will state those specifications on the sales literature.

Note that all the tools described above provide an angular measurement in relation to earths' gravity. This simplifies things for the purpose of rigging usually, but it is still possible to use a *relative measuring protractor* to determine angles (Figure 3-7) in many cases. The difference in the application is that the relative indicating protractor needs another surface from which it can be referenced. This can be difficult on an airplane because the surfaces are usually curved. It can also lead to an accumulation of errors. The rudder is one place where some sort of relative measurement must be made since a leveling type of device is not going to work in this orientation.

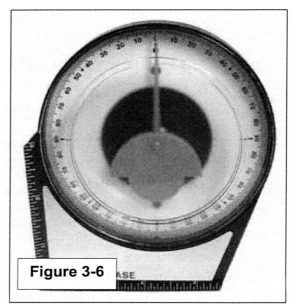

Figure 3-6

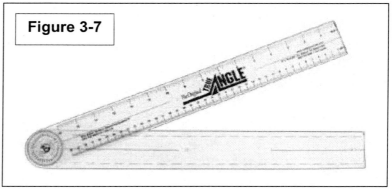

Figure 3-7

Cross-Angle or Cross-Axis Error

Cross-angle error occurs in measuring control deflections with a clinometer, where the hinge axis is not perpendicular to gravity. In Figure 3-8, the clinometer on the control surface is not vertical because of dihedral, but the bubble vial was calibrated during manufacture of the instrument on the assumption that the working surface of the clinometer would be level on a plane that is perpendicular to its measuring plane. Note that the propeller protractor in Figure 3-4 has a cross-level that folds out to be perpendicular to the axis of measurement, to enable the operator to set the protractor perpendicular to the plane of measurement. In most cases in airplane rigging the cross-angle error is very small and can be ignored, but one should be aware of errors that can accumulate because of it.

Figure 3-8

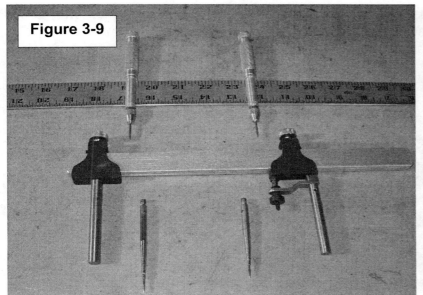

Figure 3-9

Trammel Bar

A long trammel bar is useful for squaring up the various parts of the aircraft (wing to the fuselage for example), and is more accurate and repeatable than using a tape measure (Figure 3-9). Two or three aluminum extrusions from the hardware store about eight feet long and clamped together make a long trammel bar (generally requires two people to handle).

Plumb Bob

The definition of *plumb* is to be vertical (parallel to gravity). Plumb bobs (Figure 3-10) are occasionally used to level the fuselage on all axes simultaneously by suspending it from the cabin ceiling to a known point in the fuselage. This is more common on high wing aircraft or aircraft with large cabins. The plumb bob is faster than using a level but has less repeatability than other systems for a number of reasons. If the rigger decides it is an appropriate alternative, the plumb bob setup as a whole should at least be checked against a fuselage established with levels before relying on it. The plumb bob is especially useful in rigging the landing gear.

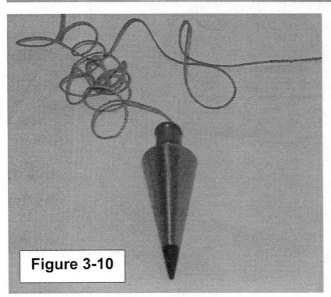

Figure 3-10

Adapters and Aircraft Specific Rigging Tools

Adapters in this context are pieces of wood, metal, or whatever, that allow one to use a level, protractor, or other rigging tool on a part of the airplane to make a specific measurement. Two common adapters are the dihedral board and incidence board, discussed shortly. They are both used with levels to make measurements.

A level may be used to make very accurate angular measurements by changing the height of one end to produce a known angle. Part of the reason for using a level to make angular measurements is that it spans a large distance, giving a more accurate indication of the condition of long surfaces (averaging out small surface deviations). The level is a very accurate instrument that is relatively inexpensive.

Because the parts being measured are the ones being rotated, the setup is designed so that the bubble is in the center when the thing being measured is at the correct angle. The difference in height at one end of the level is calculated using trigonometry (Figure 3-11). With a known angle (c) in degrees, and the length of the level (A), (B) is calculated with;

$\tan(c) \times A = B$

Be sure the calculator is set to display degrees instead of radians. With the adapter placed between the level and the surface being measured, the bubble is centered when angle "c" is correct. More trigonometric functions are given in Appendix A, Math for Rigging.

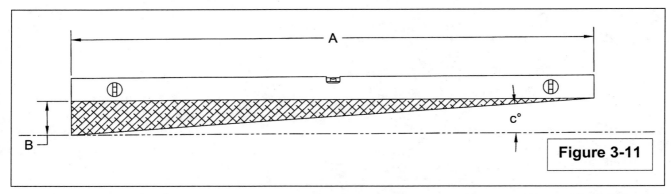

Figure 3-11

Adapters like the one shown in Figure 3-11 and the following sections are generally fabricated from cabinet grade plywood to prevent warping. If they are expected to be used a great deal, they may be varnished to prevent the grain from raising. The sharp edges should be chamfered (removed) with sandpaper as they are typically the most delicate spot.

Dihedral Board

The adapter in Figure 3-12 is a dihedral board. Dihedral is usually measured on the top or bottom of the main spar, although another location may be specified by the instructions or plans (Chapter 1). Checking in several places along the spar on each wing and dividing the difference will result in the best accuracy.

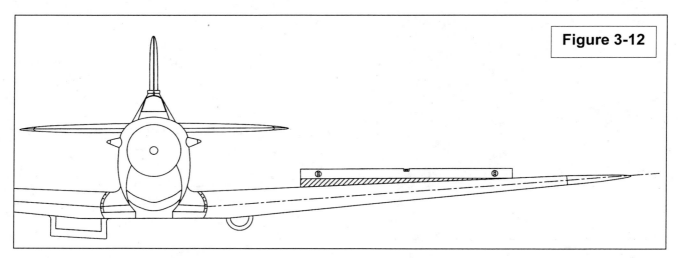

Figure 3-12

Incidence Board

Incidence boards are used in conjunction with a level to set the incidence of the wing and horizontal stabilizer. The vertical stabilizer doesn't allow the use of a level so other methods will be presented for that. Incidence boards are arranged chordwise on the wing or horizontal stabilizer to permit the incidence to be set to the correct angle (assuming the longitudinal axis of the aircraft has previously been leveled). The bottom of the incidence board must in someway conform to the curved surface of the top of the wing section, however it is not necessary to use a curved surface on the incidence board. Two different incidence boards are illustrated in Figure 3-13.

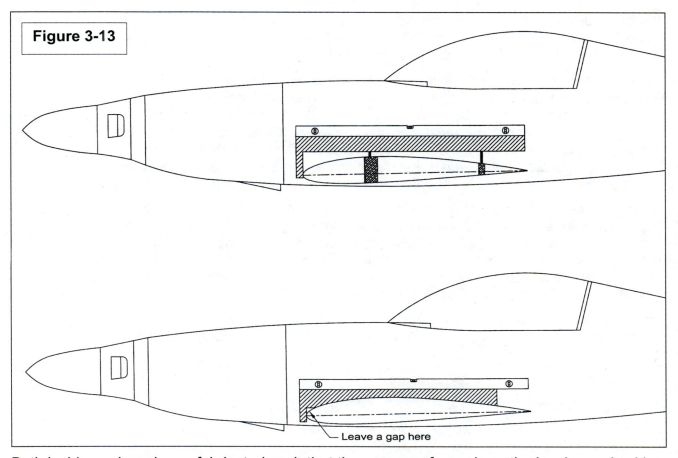

Figure 3-13

Leave a gap here

Both incidence boards are fabricated such that the upper surface where the level goes is either parallel to the wing section chord line (bubble in the middle at 0° incidence), or, they are designed so that they are at an angle to the chord line by the desired amount of incidence (level bubble in the middle when the wing is set to the correct incidence). If there is the intention to experiment with the incidence, design the incidence board so that the upper surface is parallel to the *chord line* or *zero lift line* of the wing section (see Chapter 1 for definitions). This makes it easier to do the math required to use shims under one end of the level to get the desired incidence.

Both incidence boards are designed to 'hook onto' the leading edge of the wing while they are being used. This ensures accurate and consistent results and is necessary unless there are marks made on the wing to allow the incidence board to be removed and replaced in the exact same position.

The incidence board in the upper half of the illustration uses two bolts (with rounded contact surfaces). The two bolts are the only thing in contact with the wing. They are adjustable to allow ease of fabrication. This type of incidence board is good where there are large surface imperfections in the wing (a fabric covered wing for example) as the bolts are designed to only contact the spars. Most wings are comprised of two or more spars and the skin tends to follow the spars very closely even in the presence of surface imperfections elsewhere, so providing a good indication of the actual geometry of the wing section.

The incidence board illustrated in the lower half of Figure 3-13 will produce the greatest accuracy on a wing which is accurately built and has few surface imperfections, however it harder to fabricate. It is cut from 3/4" plywood using the rib drawings to produce the desired curve (the rib drawings often don't show the wing skin so the cutout on the bottom of the incidence board must be made larger by the thickness of the wing skin, or it won't conform to the surface accurately). To keep the plywood incidence board stiff and straight, screw an aluminum channel or 'L' extrusion to the top surface where the level goes (being careful not to distort the aluminum straight-edge where the level lays).

To eliminate any surface variations close to the leading edge, remove that portion of the incidence board within a few inches of the leading edge, leaving only a small vertical portion that contacts the foremost part of the leading edge. Where an incidence board is used on a part of a wing section with a movable trailing edge surface, more accuracy may be gained by not attempting to span the control surface. This helps to eliminate variations caused by slight deviations from the control surface installation. On the other hand, extending the incidence board to the trailing edge will allow it to be used as a neutral board (discussed later).

Because many wings have tapered planforms, it may be necessary to fabricate several incidence boards to fit on the various parts of the wing. Even if the wash-out is not adjustable, the greatest accuracy and performance will be had by making several incidence measurements at different spanwise locations and averaging the readings. The incidence board in the upper part of Figure 3-13 may have room for an extra bolt and a place to attach a leading edge hook, allowing it to be used on several spanwise locations.

Many wings have a flat bottom from which incidence measurements may be made. An incidence board in this case would look like the dihedral board discussed previously.

Accuracy and repeatability in this system will have a large effect on aircraft performance because the wing reacts the largest amount of total aerodynamic force on an aircraft.

Throw Board

Control throw boards (Figure 3-14) may be substituted for angular measurement devices however it is generally easier to use the protractors discussed earlier. Throw boards are fabricated in a similar manner to the incidence boards. The rudder does not lend itself to using protractors because of the vertical orientation.

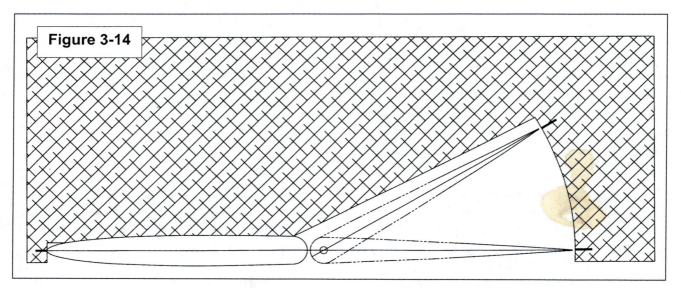

Figure 3-14

Neutral Board

A neutral board is used to hold a control surface to the neutral position during rigging of the control systems, or, to set the flaps in their neutral position (Figure 3-15).

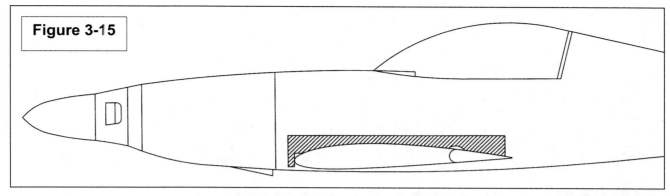

Figure 3-15

Some surfaces like ailerons are surrounded on both sides by the airfoil that makes up the wing section and so are easy to align with the neutral position (trailing edges flush as wing A in Figure 3-16). No neutral boards are necessary in this case if the trailing edge is fairly straight (however some method may be required to hold the surface in the correct position while rigging, see Chapter 5). Wing B may not have a good reference. It is adjacent to the flaps on one end (which are adjustable in themselves), and the small area at the tip tends not to be a good representation of the trailing edge because it is short and may have significant deviation from the basic wing section due to tolerances in fabrication.

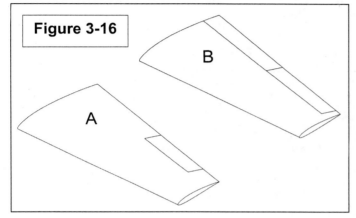

Figure 3-16

Accuracy in setting the neutral position of the flaps will have a large effect on aircraft performance because even small deviations will create a significant difference in lift and drag between the two sides, producing both rolling and yawing moments. For a wing which has a flat underside, a simple straightedge arranged chordwise along the underside of the wing will provide a good neutral location.

Just as in fabricating the other types of boards, leave a gap on the leading edge radius and near the hinge point, and shorten the board to fall slightly short of the trailing edge.

The rudder and elevator often run full span of the surface and have no convenient reference with which to align the trailing edges (left side of Figure 3-17). Even with aerodynamic balance horns as in the right side of Figure 3-17 that may be aligned with the leading edge, the accuracy is most questionable because of fabricating tolerances and the short length of the leading edge of the horn.

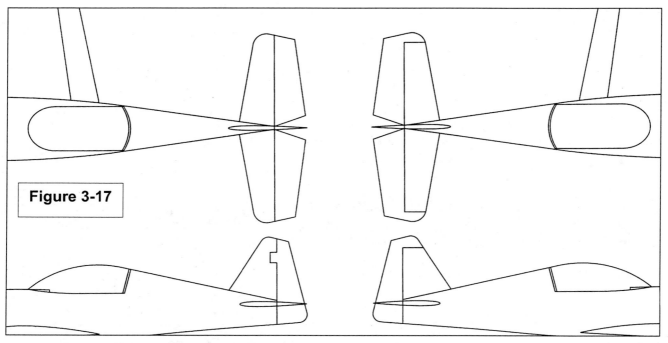

Figure 3-17

In these cases, a neutral board illustrated in Figure 3-15 will align the surfaces in their neutral position. Like the incidence boards described earlier, they are fabricated using the airfoil template included in the plans or manual for the aircraft. A simple neutral measurement for symmetrical airfoils is illustrated in Figure 3-18. When the two wood spacer blocks are of equal thickness, the control is at neutral.

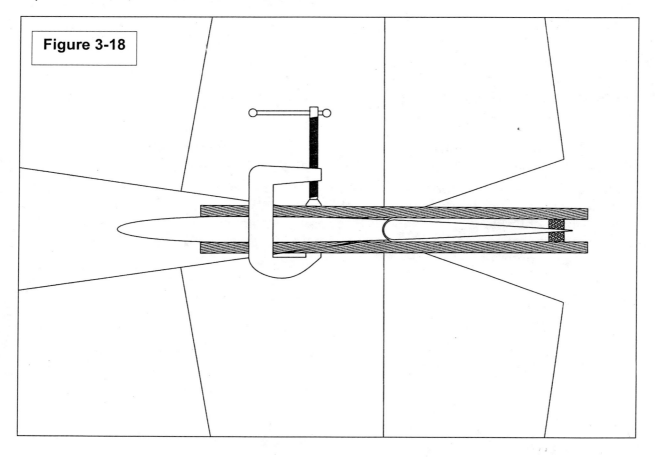

Figure 3-18

Figure 3-19

Figure 3-20

Control Locks

Control locks are needed to hold to hold the stick or pedals in their neutral position, or control surfaces in their neutral position for rigging the control system. Some pieces of wood and C-clamps often are suitable. Some aircraft are designed to accept pins that hold the control actuators in their neutral position (Figures 3-19 and 3-20), while some rigging procedures call for the control surface to be clamped in the neutral position (neutral boards are useful here). Due to the wide variation of control system design, it is up to the rigger to be imaginative on these points. This is discussed again in Chapter 5.

Thrustline Measurement Tools

Checking and setting the thrustline can be done in several ways depending on the accuracy required. Exactly where it's supposed to be is generally stated in the plans or manual, but for performance it's often a matter of experimentation, which requires a method of making repeatable measurements. A lot of engines may not have a surface from which to measure the thrust line from. The thrust line needs to be measured both in the vertical and horizontal plane. The vertical plane is most easily done with a level, and something else will be needed to set the horizontal plane. Simpler methods than given here are provided in Chapter 5, but are not as accurate or repeatable.

Vertical Plane

The most accurate and consistent method to set the vertical thrustline is with a spindle that attaches to the propeller hub as in Figure 3-21. This allows a level to be used in its most favorable orientation. The spindle must be machined perpendicular to the mating face.

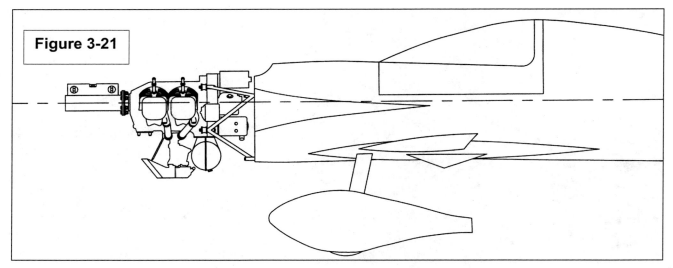

Figure 3-21

A simpler (but potentially less accurate) method of aligning the vertical thrust line is to attach a test bar to the prop hub as in Figure 3-22, and use a level. Construct the test bar by obtaining a long piece of rectangular steel or aluminum tubing that will be drilled to enable it to be bolted to the propeller hub. The longer the tube, the more accurate will be the measurements, and the tube must be straight. Tracking it like a propeller won't help if its not straight.

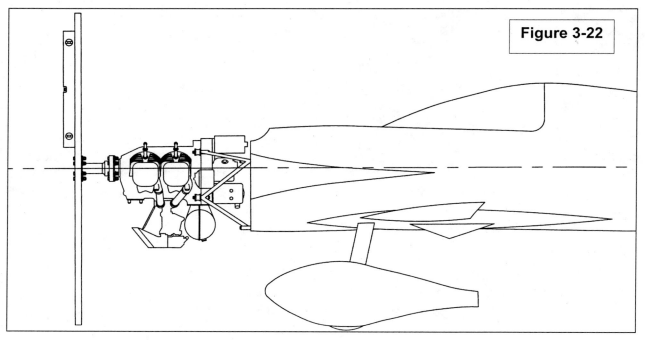

Figure 3-22

Horizontal Plane

The horizontal plane is easily set by using the bar described in the previous paragraph, but applied as in Figure 3-23. Make two marks on the tube, as far away from the center as possible, each exactly the same distance from the center of the propeller hub. Find the aircraft centerline at a convenient point some distance back from the engine. Level the bar horizontally and use a trammel bar to measure the relationship between the marks on the tube and the aircraft centerline. In the example shown in Figure 3-23, the thrust line is being made perpendicular to the aircraft centerline. If the thrustline is not perpendicular for a particular aircraft, the trigonometric functions may be used to calculate the linear difference necessary for the offset (see Appendix A). More ideas for setting the thrust line are given in Chapter 5.

The propeller may be used but it is difficult or impossible to lay a level on the surface because of the curves. In addition, many propellers are too short to get meaningful results.

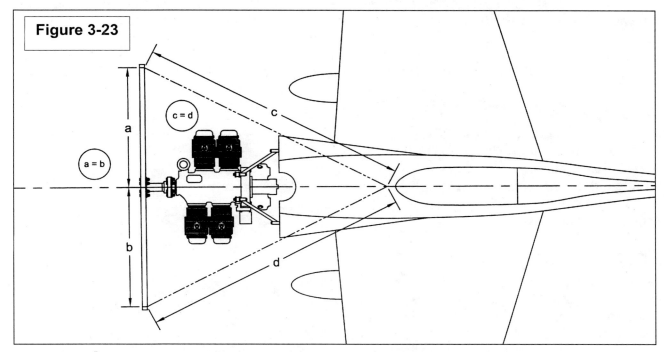

Figure 3-23

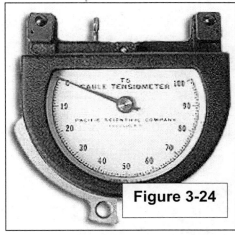

Figure 3-24

Cable Tensiometer

Cable tensiometers are relatively expensive devices (Figure 3-24). Some EAA chapters keep one for the use of its members, or tensioning may be accomplished by bringing the aircraft to a maintenance shop, or having the mechanic bring it to ones own shop. There are many different types, some with much greater accuracy than others. Most cable tensioning tolerances allow at least 10% variation which is consistent with the accuracy of the lowest cost models. Note that different cable diameters usually require different settings on the tool, or, in some cases, a different tool.

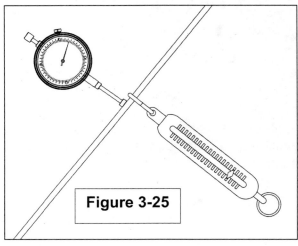

Figure 3-25

Wire Tensiometer

Wire tensiometers are used to set the wire tension on aircraft with externally braced flying surfaces (biplane wings, empennages, etc.). They are relatively expensive and tensioning may be accomplished by bringing the aircraft to a maintenance shop, or having the mechanic bring it to ones own shop. The maintenance manuals for many aircraft with flying wires do not have a specified tension, perhaps only that they make a certain noise when plucked like a guitar string. It is still important to achieve consistent tension in a circuit of wires regardless of the method used.

For lack of expensive tensioning tools, even tensions among the various wires (of the same size) can be made with a fish scale and a dial indicator or ruler (Figure 3-25). Applying a known force to the wire will produce a given deflection. Take measurements in the center of the wire. It is best to attach a short string to the scale and the flying wire to avoid scratching the wire with the scale. Some designers specify this method to determine tension and provide a deflection measurement along with the weight to be applied. Even when the actual tension is unknown, this method may be used to make all of the tensions in a circuit the same.

Streamlined Flying Wire Tool

A pair of wrenches (Figure 3-26) fabricated from phenolic, nylon, brass, or aluminum, makes turning the streamlined flying wires easier without marring or scratching them.

Torque Seal

Torque seal is a glue-like substance, applied from a tube, that hardens in air. It is used on threaded fasteners to indicate if they have rotated from a preset position (Figure 3-27 and 3-28). If the fastener were to loosen or move, the dried torque seal cracks, providing an obvious indication to the inspector. It is rugged and withstands large temperature variations and vibration if applied to a clean, degreased surface. It is available in several bright colors.

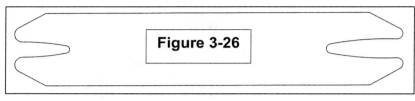

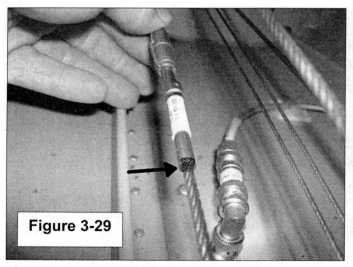

Similar to the idea of torque seal is the application of paint to swaged cable ends (Figure 3-29), to provide an effective way to determine if the terminal has slipped from the cable because of a faulty or overstressed joint.

Turnbuckle Holder

The inexpensive tool illustrated is Figure 3-30 will make adjusting turnbuckles much easier. It holds the turnbuckle ends while the turnbuckle barrel is rotated, preventing the cables from twisting. Two of them make rigging much easier.

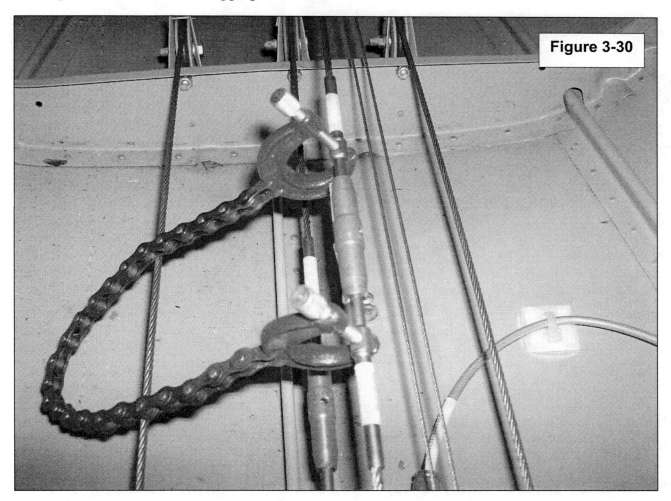

Figure 3-30

Figure 3-31

Surveying Transit and Similar Equipment for Rigging

The main flying surfaces of airplanes are sometimes rigged using a surveying transit (Figure 3-31). With two people working this is a fast method and for larger aircraft is the only method available. For a small aircraft, the consistent application of levels and careful measurements will produce the same accuracy or better, although at a slower pace. Much of the equipment used in construction site layout will produce accuracy no better than 1/8" vertical, and is generally unacceptable for airplane rigging.

Slip/Skid Indicator

The slip/skid indicator can be one of the most important instruments in determining the rigging state of the aircraft. See Chapter 2 about the relationship between the inclinometer and aircraft motion.

Some examples are shown in Figure 3-32. Most familiar are probably the turn coordinators that have the slip/skid indicator installed in them. It is two separate instruments combined into one for reasons explained in the paragraph on Turn Coordinators.

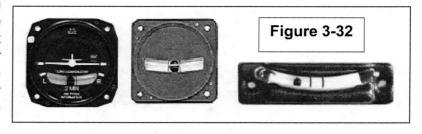

Figure 3-32

Many inexpensive slip/skid indicators have an air bubble in the vial to allow for expansion and contraction. If the bubble is big, it sometimes causes the ball to be less sensitive or slightly inaccurate when the meniscus of the fluid attaches to the ball. The better slip/skid indicators have a separate expansion chamber in the instrument case, so no bubble is visible (if a bubble appears out of nowhere, roll upside down and it will go back into its expansion chamber).

A longer slip/skid indicator generally has greater resolution, and a lesser curve (larger radius) of the vial increases sensitivity. To get the most performance from the aircraft during rigging, install a long slip/skid indicators with a large radius, as a rigging tool. When rigging is complete put a little one on for flying. If the desired inclinometer can't be located, sensitivity may be increased by installing it on a tapered adapter between the instrument panel and inclinometer as in Figure 3-33. This has the same effect as reducing the curvature (increasing radius). If too much taper is used, the action of the ball becomes non-linear and can be difficult to interpret. Instructions for setting the inclinometer are given in Chapter 5.

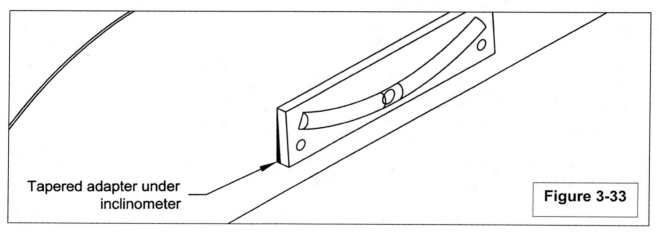

Tapered adapter under inclinometer

Figure 3-33

Yaw String

A yaw string is a good substitute for a slip/skid indicator, or can be useful for visualizing what the inclinometer is indicating. In Figure 3-34, the relationship between the inclinometer and the yaw string is indicated for a turning aircraft.

The ability of a yaw string to produce useful information varies from aircraft to aircraft depending on the airflow which impinges on the front of the canopy where the yaw string is placed. They are best suited to gliders or to aircraft whose engines' thrust does not impinge on the canopy/windshield.

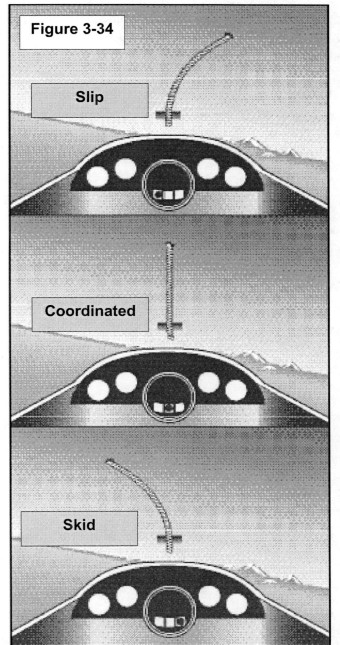

Figure 3-34

Attitude Indicator

Attitude indicators can provide some useful rigging information. Attitude indicators have a subjective resolution (index marks are typically at 5° apart for roll and pitch, at best) and they are only as good as their installation in the panel. Although the gyro in the case is self erecting and automatically orients itself to gravity, the index marks for bank angle are on the instrument case. The instrument case is adjustable in relation to the gyro by loosening the screws holding the instrument to the panel and rotating the whole instrument case around the gyro. Normally they are adjusted to provide useful roll reference to the pilot *after* an aircraft is rigged. Unless it can be run and adjusted with the aircraft leveled for rigging, it may not provide any useful information for rigging. Information given by the attitude indicator can be inferred from some combination of other instruments, but it may help in visualizing what the aircraft is doing during testing of the rig.

Turn Coordinator/Turn and Bank Indicator/Turn and Slip Indicator

The *turn coordinator/turn and bank indicator/turn and slip indicator* (left side of Figure 3-32) are rate gyros. They are designed to only provide an indication of the rate of turn (degrees per second) of an aircraft. Because of this they may provide some useful rigging information depending on the nature of the misrig of the airplane. The different types differ slightly in design and display but the differences are relatively unimportant to rigging. They all give meaningful information about the rate of turn *only* if the ball of the inclinometer is centered (which is why the two instruments are commonly combined).

NOTE

For IFR operations, a turn coordinator (rate gyro) is no longer required equipment if a second attitude indicator with an independent power source is installed. At the time of this writing, this has not been written into law (14CFR91.205), but refer to Advisory Circular 91-75. A slip/skid indicator is still required for IFR operations.

They are limited to small bank angles (bank angles that make a standard 3° per second turn rate used in IFR operations). The bank angle required for a standard rate turn varies with the airspeed of the aircraft but IFR operations limit the angle of bank for maneuvering to 30°.

The rate gyros can be useful for rigging because they will indicate small turn rates, which otherwise may be difficult to detect except by piloting very carefully and watching the compass or heading indicator. This instrument is adjustable in the panel like the attitude indicator and must be set so the ball is in the middle with the aircraft leveled laterally.

Compass and Heading Indicator

The compass and/or heading indicator may be used as a reference for straight flight and detecting slow turns. The magnetic compass can be difficult to use in turbulence, and magnetic dip tends to magnify small changes. Precession from magnetic dip will be minimized by flying east or west.

Chapter 4
Factors in Rigging

Control Friction

Accurate rigging requires that the control system be free of excessive friction. Friction in the control system can either give the illusion of a rigging problem, or, mask a rigging problem. In really bad cases, friction masks the feedback to the pilot from the airloads on the flight controls, making the airplane unpleasant to operate. Frequently when it is stated that an aircraft has flown hands off the first time, it is often due to friction in the controls. To put it another way, the pilot was able to move the controls to the desired position and they would stay there, regardless of the airflow imposed on the control surfaces trying to align them with the relative wind. It is impossible to eliminate all friction, however, there shouldn't be so much friction in a control circuit that the airflow at low flight speeds won't make the control surfaces deflect or realign with the local relative wind. An increase in control friction also makes an airplane seem more stable because the control surface doesn't float or align itself with the relative wind.

For controllability in small airplanes, four pounds of friction in the rudder, two pounds of friction in the elevator, and one pound of friction in the ailerons are considered an upper limit. It is desired for accurate rigging and pleasant flying that the friction be much less than this.

Excessive cable tension increases control friction. See the paragraph on Cable Tensions.

Fairleads used to change cable direction are a common cause of excessive friction, especially if the fairlead has a deep groove worn it. The only thing that can be done, short of redesigning with pulleys, is replace or rotate the fairleads and lubricate the inside to help reduce the friction. 14CFR23 doesn't allow the use of fairleads to change cable direction by more than three degrees, however, many experimental aircraft use fairleads for larger direction changes.

Control pulleys of insufficient diameter vs. amount of cable direction change vs. cable diameter are another source of *apparent* friction. A cable of a certain diameter, forced to bend around a too small radius is constantly working against itself because of the stiffness of the cable. To change a cables' direction by more than fifteen degrees requires a 3.5" diameter pulley for 1/8" diameter control cable.

Pulleys made for primary flight controls rotate on sealed ball bearings. Some homebuilts may have less expensive pulleys using bushings that need to be lubricated frequently.

Improperly designed control linkage geometry will result in the cables changing tension as the controls are deflected to different positions. This will be an apparent source of friction.

Hinge points without ball bearings are another cause of friction. The amount of friction that exists between two moving components is a function of the types of materials and the pressure exerted to force them together (note that the amount of surface area of the contacting parts does not influence the force required to overcome friction). The force required to overcome friction increases proportionally with the force exerted between the parts. The forces that apply pressure to the components in a control system are cable tension, airloads, and acceleration (pulling g's). Even when the system seems free of friction on the ground, flight loads on the control surfaces will exert forces on the hinge points that will increase the friction considerably.

Certain materials work together better than others to reduce friction (bronze and steel, aluminum and steel). Steel that rubs on steel is constantly creating tiny burrs at the contact area which 'grabs' under a load (galling). In some well used systems it may be necessary to remove the hinge bolt, clean/deburr the contact areas, and/or install a new bolt. In any case, hinge points that use plain bearings (bushings) or no bearings at all should be lubricated prior to adjusting rigging and prior to testing the rig in flight.

Thrust washers may be employed on some hinge points to help reduce friction and galling, check to make sure they are present when necessary and in good condition.

Friction is greater when an object is first started to move (static friction), as opposed to once it's moving (rolling friction). In terms of control movement this is referred to as *breakout force*. High breakout forces make small control inputs and accurate trimming difficult or impossible. Monoball bearings have a higher breakout force than ball bearings, whether used as hinge points or on rod ends.

A bent or warped structure/control surface may also be responsible for friction if the hinge axes are no longer aligned. This is commonly caused by improperly adjusted flying wires. Adjusting flying wires is discussed later.

Control Cable Tension

An aircraft built from plans will often have no specified cable tension for a closed loop cable system. Rudder cables are frequently tensioned by springs (open loop control system) and require no tension adjustment. Required cable tension is difficult to calculate and is a function of the both the expansion and contraction rate of the airframe materials, and, the ability of the control system to resist the strain imposed by the pretensioned cables. Excessive cable tension increases control friction. Some guidance is provided and it is up to the builder/rigger to determine what is correct.

An aircraft fuselage which increases in temperature is going to expand in length and width, sometimes at a faster rate than the cables expand. This will result in an increase in cable tension if the cables were initially adjusted at a lower temperature. In contrast, an airframe which undergoes a drop in temperature is going to contract faster than the cables and will leave the tension less than what was originally set at some higher temperature. The amount of change in cable tension can be quite substantial. Aluminum aircraft expand and contract more than other airframe types, and do so at a faster rate than the steel control cables. A small two seat aluminum airplane will have noticeably looser cables in the winters of the northern US if the cables were adjusted at the typical 70°F (large aircraft have automatic tensioners). Cable tension figures given in maintenance documentation will include the temperature at which the cables should be adjusted. In reviewing the maintenance instructions for a variety of small aircraft of different types of construction, the 1/8" primary control cables will vary in tension approximately 2-5 pounds per 10° F change in temperature.

Lacking proper tensioning data is going to require that the cable tension be regularly checked for *correctness* (a partially subjective evaluation for suitable tension) until some value is found that works at all the expected temperatures. The longer the cable or the bigger the aircraft, the more critical the tension is because of temperature changes.

In evaluating *correctness*, use the following points as a guide;

- Excessive cable tension results in increased and rapid pully/bearing wear and increased control friction, especially where a cable rubs on a fairlead or where hinge points are not riding on ball bearings.
- Too little cable tension results in sloppy controls (relatively larger control inputs to effect a desired flightpath change) and sometimes odd flying characteristics. It may also

cause control interference. Too little cable tension may result in the sagging cables rubbing or snagging on other objects. Loose control systems negatively affect an aircrafts resistance to flutter.

- It is possible to have insufficient control authority at high speeds due to loose cables (the airloads on the control surface cause them to be driven toward neutral). Elevator controls which are floating freely slightly because of loose cables affect airplane longitudinal stability in some flight conditions.

- In maneuvering and turbulence, the loose cables oscillating up and down may be transmitted back into the control wheel or stick and the control surfaces to which the cable is attached, a highly annoying and fatiguing way to fly. In addition, the oscillating cables may be acting on the surface to which they are attached to produce lots of little deflections that increase drag, and the control surface to which they are attached may be floating freely. Where the oscillating control cables are a problem and the tension seems correct, additional fairleads should be installed.

Most general aviation aircraft with cable controls do not exceed 50 LBS of tension on the primary flight controls, with 20-35 LBS being most common. Open-loop rudder cable systems tend to be less than this because they are held with relatively weak spring tension. Because of the long lengths and insufficient tension on open-loop rudder cable systems to prevent them from oscillating during aircraft movement, they need to be contained in fairleads more frequently along their length. In looking at the maintenance instructions for a variety of small aircraft, it is seen that the elevator cables usually have more tension than the aileron cables, often twice as much.

In no case should the control cable system stretch or shorten by more than 25% under a load. The percentage of stretch is calculated from;

$D = 100 \times a / A$

where;

D = Cable stretch, percentage

a = movement of cockpit control with control surface clamped in position (inches)

A = available movement of cockpit control when control surface is free (inches)

For a, the deflection is measured using a force given in Table 4-1.

Table 4-1

Control	Pilot Force, Pounds	Application
Pitch	88	Push or Pull
Roll	44	Sideways
Yaw	100	Forward Press on Pedal

NOTE

The forces to be applied in Table 4-1 are measured at the top of the stick, or at the outer edge of the control wheel, as appropriate.

Free play in the controls, whether caused by insufficient cable tension or sloppy hinges, are conducive to flutter. Flutter can occur on any aircraft at any speed if there is enough free-play in the controls. Trim tabs and ailerons are particularly dangerous in this regard (discussed in later paragraphs).

Adjusting Cable Tensions

On a closed-loop flight control system (see Chapter 1), a single cable adjusted to the proper tension anywhere in the system should theoretically put the same load on the other cables in the system. This is rarely the case on many cable control systems because of friction in the controls and the weight of the control surface applying a tension to one circuit of cable. It is possible in many cable systems to observe a very large difference in cable tension, in a closed loop system, between the two cables for a single control (all cables being identical in routing and function).

To obtain an even tension and distribution of loads on the control system, it is necessary to support the control surface being rigged so that its' weight doesn't affect the tension readings. It is also necessary to 'twang' the cable with a finger and/or wiggle the controls around after making a tension adjustment and before taking a measurement, to force things into equilibrium (this is because of friction in the controls).

Some aircraft maintenance manuals will specify a single point from which to take tension measurements, rather than specify that the control surface must be relieved of weight for cable rigging. These manuals will also state to ignore the tension on the other cable.

It's good practice when rigging the cable control systems to inspect the pulleys for freedom of movement while the cable tension is off. In addition, rotating the pulleys 180° under the cable will promote even wear. See also FAA Advisory Circular 43.13-1B Aircraft Inspection and Repair on inspecting pulleys for wear.

It is suggested that new aircraft have their cables tensioned initially about 25% too high, moved around a bit, then loosened and retensioned to the correct tension. This will help keep the cables at the correct tension later.

Aircraft which are operated frequently from unimproved airstrips will need more frequent rigging of the rudder/tailwheel cable tensions. The tailwheel cables need considerably less tension than the rudder cables in a closed circuit rudder system. While the rudder may be rigged to 30-40 pounds of tension (in a closed circuit system), the tailwheel steering cables may only require 5-10 pounds.

Trim systems also require much less tension than primary flight controls, usually twenty pounds or less. Trim cables are usually of a smaller diameter than primary flight control cables.

Control Wear or Sloppiness

Loose or sloppy controls, either caused by worn hinge points or loose cables, can affect the interpretation of both the rigging instruments on the ground and the flying characteristics of the aircraft. When in the air, loose controls may continuously try to orient themselves with the local pressure distribution, never getting here or there in certain flight regimes. The effect is subtle but is most easily recognized by the inability to *precisely set a control at a particular deflection to produce an exact desired result in the flight path or energy level of the aircraft*, in one or more flight conditions. Piloting experience and/or familiarity with a particular aircraft plays a role in being able to detect these subtle differences. A phenomenon called *snaking* where the airplane is continuously seeking equilibrium about the vertical axis can sometimes be caused by very sloppy or loose controls, but more frequently is caused by a more fundamental stability problem. In any case, sloppy controls will not produce the desired performance, and will affect both the in-flight interpretation of rigging and the interpretation of the rigging measurements on the ground..

Loose or sloppy controls may instigate or sustain flutter. Loose trim tabs are the leading cause of flutter incidents in small aircraft. A completely disconnected trim tab may cause serious flutter at speeds as low as 50 KTS, maybe even less. The total free play at the tab trailing edge should be less than 2.5 percent of the tab chord (Figure 4-1). The total free play at the aileron trailing edge should be less than 2.5 percent of its chord when the other aileron is clamped solidly.

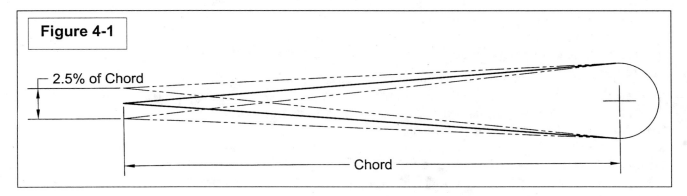

Figure 4-1

Control Stops

Certificated aircraft are designed with adjustable stops on all the controls, but some experimental aircraft designs do not provide for control stops. Adjustable rudder control stops are illustrated in Figure 4-2.

Control stops are important from both a safety and long-term wear perspective. Control stops should be installed that contact both the movable surface, and the actuator (stick, pedals). Without control stops in both places, the full deflection of the stick or pedals will place a considerable load on the control system itself, including cables, bellcranks, push-pull rods, pulleys, etc.. At the very least a great deal of wear and 'stretch' will occur as the controls are moved to their extreme deflections. At worst, a really hard control deflection is going to break part of that control circuit, or even just cause something in the system to go over center and stay that way (like a bellcrank or control horn on a control surface). Aerodynamic phenomenon can also drive a control surface to extreme deflection, causing problems with the mechanical linkage and obviously the flight characteristics of the aircraft. A windy day parked on the ground without control locks will also cause the flying surfaces to be repeatedly jammed to the limits of movement.

Figure 4-2

When rigging a control circuit, the maximum deflection of the control surface is set by the control stops at the control surface. Then the stops at the actuator (stick or pedals) are adjusted so they contact slightly *after* the control surface contacts its stop. Because of some flexibility that exists in all control systems, little force is required to continue moving the actuator to contact its' stops even after the control surface has contacted its' stop.

The amount of deflection that an actuator is allowed after the control surface has hit its' stop depends on the type of control system and the expected working speed of the aircraft.

In a rigid control system (push-pull tubes without much linkage for example) the gap between the actuator and its control stop may be very small, less than .015", when the control surface is contacting its' stop. On a more flexible system (one that includes cables, pulleys, and bellcranks), it may be necessary to deflect the actuator a 1/32 inch or more to obtain the required travel of the control surface, at the working speed of the aircraft (discussed next).

The expected working speed of an aircraft plays a large part in determining the difference between the control stops, because higher velocities will work on the control surfaces to deflect them away from their stops (due to the flexibility in the control system). Remember from Chapter 2 that the pressure of the air hitting an object increase as the square of the velocity, so even a twenty knot change in velocity is going to create a large change in the force acting on the control surface. Aircraft certified to 14CFR23 require that the control surfaces move at least 10% of their original deflection at the maximum diving speed/load factor of the aircraft (see the next section about this).

On some aircraft, the control stops are solid pieces of metal that are ground and filed away to provide the proper settings. This is an easy alternative to the adjustable stops illustrated in Figure 4-2, even when material must be added to provide the correct setting.

The adjustable stops illustrated in Figure 4-2 are locked with both a jam nut and safety wire. The safety wire is essential, there have been problems (and fatalities) on certificated aircraft with adjustable stops that were secured only with a jam nut. The stops loosened and backed out under vibration, restricting control movement. Regardless of the control stop design, it is imperative that they not be allowed to accidentally move from their rigged position.

Control System Operational Tests Per FAR 23.683

Notwithstanding the cable tension or flutter requirements, some basic control authority must be available at the limit loads of the aircraft. FAA Advisory Circular 23.683-1, 23-19, and 14CFR23 provide instructions on performing operational tests using realistic loads on the controls and control surfaces. Appendix A to 14CFR23 provides simplified design criteria to determine the expected loads on the control surface. All of these documents are available free from the FAA website. Although the specifications in those documents are only required on certificated aircraft, they provide a realistic prediction of what an airplane will expect to see in service.

Gap Seals

Gap seals are often used on control surfaces to prevent the leakage of high pressure air on one side to the low pressure air on the other side. This can significantly increase the control power and reduce drag.

Replacing gap seals on aircraft so equipped must be approached with caution. There is a real possibility of it fowling a primary flight control if the seal comes loose during flight. Installing gap seals requires careful forethought as to making it failsafe. Many gap seals depend on adhesives whose performance varies greatly with the environmental conditions as well as the applicators skill.

Two examples of gap seals are illustrated in Figure 4-3. The seal in the upper part of the figure is a piece of fabric or tape with one edge bonded to the fixed portion and the other bonded to the movable portion (the bond is at the circles). The seal in the lower part of the figure is a foam seal similar to weatherstripping. This seal requires the leading edge radius of the control surface to be nearly constant and concentric to its' hinge axis.

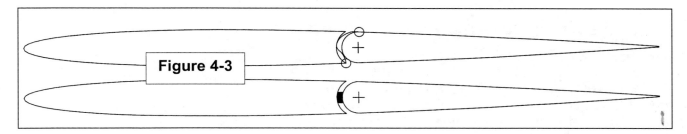

Figure 4-3

General Control Checks

Generally one will walk away from an engine failure in a single-engine airplane (in VFR in the daytime anyway), but a stuck elevator or ailerons is going to be tough. Before flying the rig, check the controls for proper operation and potential fouling.

The control surfaces should move in the proper directions relative to the actuator (stick or pedals). There has been more than one accident caused by backwards rigging, and it's not hard to do in some aircraft.

Check that aerodynamic or acceleration loads aren't going to move anything around and make the controls jam up, especially where clearances are small. On some homebuilts, a very large temperature change may cause things to bind up.

If it has just had a lot of work on it, make a check for loose tools, hardware, and missing safety wire/cotter pins/nuts/bolts/pulley guards. Cotter pins used as pulley guards are shown in Figure 4-4. They prevent the cables from getting off the pulleys and fouling the controls. A lot of airplanes have accumulated a lot of loose hardware under the floorboards over the years, but it just takes one piece to float up and stick in something.

The controls should have no resistance in their movement. Watch out for a particular point in the deflection where there is a slight change in friction or resistance.

There should be no rubbing/chafing noises, or much any noise for that matter, as the controls are moved through their deflection. An autopilot installation where the control cables are wrapped around capstans might make a little noise, and provide a *slight* resistance to the controls. Phenolics, teflon, and nylon are employed to prevent cables from rubbing against the rest of the structure and fraying the cable. Where this material is installed, it may be possible to hear the cable rubbing occasionally during normal control operation (without the engine running).

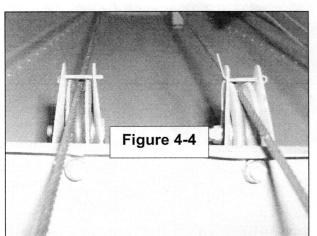

Figure 4-4

There should be a definite halt to control movements where the control/actuator hits its stop.

The rudder and elevator must not be able to touch each other under any combination of control deflections.

The last four checks should be performed both

on the outside and from the inside of the airplane.

While in the aircraft, ensure that the deflection of one control axis does not interfere in any way with the operation of the other control axes or secondary flight controls.

Hardware Encountered in Rigging

FAA <u>Advisory Circular 43.13-1B Aircraft Inspection and Repair</u> provides a great deal of information on control cable systems and safetying. Anyone involved with the construction or maintenance of aircraft should have a copy of this manual (it is very inexpensive). It is sold by most book dealers that distribute aviation technical books or it can be downloaded from the FAA's website.

Rod Ends

There are several factors to consider when working with rod ends.

The threaded portions must engage the actuator to a depth at least equal to the diameter of the threads, to develop the rated strength (A>B in Figure 4-5).

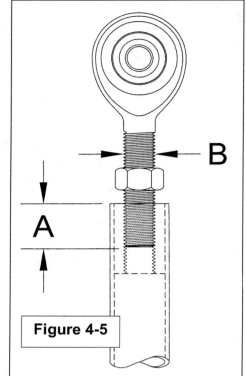

Figure 4-5

Female threaded rod ends provide a witness hole (Figure 4-6). By putting a small piece of wire or something into the witness hole it can be determined if the threads are sufficiently engaged. If the wire goes in too far, the threaded portion needs to be turned in more.

Threaded rod ends are used with a jam nut (shown loose in Figure 4-5 and 4-6).

Rod ends which are attached to something that is only on one side of the rod end, need a large washer on the other side to prevent the rod end from separating in the event the bearing fails (Figure 4-7).

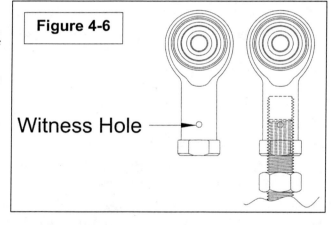

Figure 4-6

Witness Hole

Rod ends are typically used on control pushrods where the geometry of the linkage changes on several axes simultaneously. The allowable movement of the rod end on ay axis must not be exceeded at the extreme deflections of the control. This can be determined by attempting to twist the rod end (around the longitudinal axis of the pushrod) with the control deflected to both extremes. It should rotate a little. If it cannot be twisted back and forth a bit, it's either exceeding the limits of its' movement, or it may be rubbing on the control horn or washer (right side of Figure 4-8). If it's only rubbing on something, a common fix is to add small washers or spacers (the washers or spacers should have the same outside diameter as the

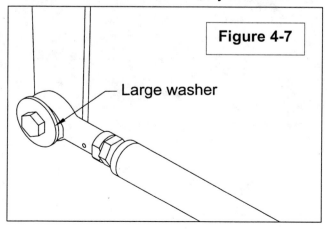

Figure 4-7

Large washer

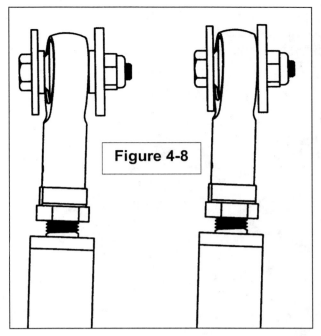

Figure 4-8

flat spot on the ball of the rod end, where the bolt goes through) and a longer bolt as in the left side of Figure 4-8. If it's truly exceeded its freedom of movement at the extreme deflections, the system should be redesigned.

If a force is applied to a control surface it is possible that the rod end won't rotate readily, because the actuator handle (stick, wheel, or pedals) is hitting the control stop before the control surface hits its' control stop. When the control actuator hits the stops before the control surface, a heavy pressure on the control surface applies pressure on the rod end bearing internally, causing the rod end to bind up and resist rotation. This is indicative of a control stop adjustment error, rather than a rod end problem. But. many low-speed aircraft have the aileron control stops at the bellcrank in the wing, and a force applied to the aileron will also bind up the rod end even if the stops are adjusted correctly. If the rod end is used in a system with an irreversible actuator (flaps for example), binding of the rod end internally may occur if the flaps are jamming against a stop, rather than being held by the actuator.

Rod ends should rotate slightly and swivel freely, but otherwise have no free play or looseness in their bearing(s). Some rod ends have grease fittings while others are sealed. Most rod ends used in primary control systems are composed of multiple ball bearings internally and are identified with a milspec number. Less expensive rod ends are of a monoball construction, where the spherical center piece of the rod end bearing rides directly on a solid low-friction liner pressed into the outer race (rod-end housing). These rod ends typically have a shorter life and higher breakout force (starting friction), especially when under a load.

When adjusting rod ends, do not turn the rod end by sticking something in the ball hole for leverage. This may cause permanent binding or damage to the rod-end. Use a wrench on the flats at the base of the rod end.

Turnbuckles

Turnbuckles are delicate and easily bent when working with cable systems. If a turnbuckle gets bent, it's safer to replace it rather than bend it back.

It is required that turnbuckles have no more than three threads exposed, to develop the full strength of the unit (Figure 4-9). Some have witness holes, similar to the rod ends. See AC43.13-1B for safety wire instructions.

Turnbuckles are made of a brass material that is relatively weak compared to steel and so they require extra care. While they have good resistance to corrosion from normal environmental factors, they will corrode extremely rapidly in the presence of certain chemicals and metals.

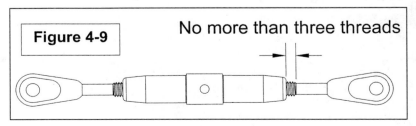

Figure 4-9 — No more than three threads

Turnbuckles should not come within two inches of fairleads or other cable guides when the controls are fully deflected. One reason is this; if the cable becomes frayed over time it often occurs near the swage that attaches the turnbuckle to the cable. As the separated strands unwind from the cable over time, they travel backwards along the length of the cable. When they encounter the fairlead or other guide, the strand(s) entangle there and create a control jam. This does not always happen with warning.

Clevis Forks

Steel clevis forks used on powerplant control systems (Figure 4-10) must be seated to a depth greater than their diameter (same as in Figure 4-5, A>B), to develop their full strength. If a witness hole is provided, go by that.

Steel clevis forks used on wing-strut ends (right side of Figure 4-11) are treated differently because the loads may not be exactly under tension or compression, there may be a bending moment imposed if everything is not exactly in a straight line along the strut. It is necessary to consider the strength of the threaded barrel that is welded in the strut end (left side of Figure 4-11) and how much eccentricity is allowed to be applied to the barrel when the clevis fork/strut is subject to a bending load. It is *safe* to assume that the clevis fork should be seated to a minimum depth approximately equal to the length of the barrel (welded-in female portion, Dimension A in Figure 4-12), to develop their full strength in all directions.

Figure 4-10

Figure 4-11

NOTE

The threaded fasteners used on aircraft structures are formed by rolling the threads rather than cutting (like taps and dies). This is the strongest method of producing threads. Threads formed by cutting are weaker, less precise, and require significantly more engagement to develop the full tensile or compressive strength. The Machinery's Handbook indicates that as little as 10% of the threads in a connection are in contact with each other under a light load, for average threads formed by cutting.

If a witness hole is provided, go by that. Usually a witness hole is not provided here, and it will be necessary to disassemble and measure the threads on the fork, and in the barrel, to determine the engagement once everything is assembled.

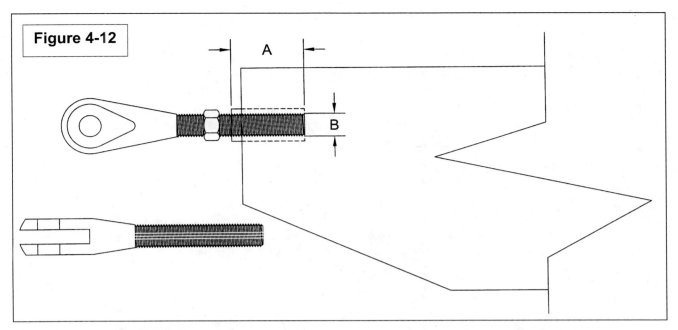

Figure 4-12

Flying Wires/Tie Rods

Figure 4-13

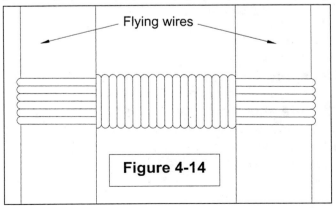

Figure 4-14

Flying wires used to brace the aircraft structure are made of solid steel or stainless steel, but in a few cases are made up of stranded cable (like that used in control systems) to save money. It is desired that an aircraft whose primary structure is braced by wires have correct tension of those wires. While some aircraft maintenance manuals will state a numerical value for tension, other manuals will say only that the wire should have a certain noise when it is plucked like a guitar string, or, that the tension should be the lowest possible without suffering from vibration in flight. It is still desirable on those aircraft, whose manuals don't specify a tension, that the opposing wires all have the same tension. The wires should not vibrate at low frequencies or they will suffer fatigue failure. Increasing tension will probably eliminate the vibration (sometimes tension must be decreased to eliminate vibration). Tools for setting the correct tensions are discussed in Chapter 3. Where wires touch each other, some anti-chafe material should be used between them. They are often secured to each other with a wooden dowel at the mid-span as in Figure 4-13 (called a javelin, birdie, bodkin, etc.), to prevent vibration that will result in metal fatigue. In addition, the closely spaced pairs of wires are sometimes laced together with heavy string at various points along their length (Figure 4-14).

Solid wires may be round or streamlined in cross-section. When assembling the ends onto the wire, apply some grease or anti-sieze compound to the threaded portion to prevent corrosion and make adjustment easier. This is necessary on stainless wires to prevent galling of the threads when tightening. It is customary that the end with the right-hand threads should either be down or forward depending on the orientation of the wire, to make it easier for the person doing the adjusting. To adjust them it is necessary to apply rotational pressure evenly to the two ends. If the two ends can't be reached simultaneously, turn each end in small increments, alternating between the two ends so as not to put a permanent twist in the wire.

The wire should make a straight line into the fitting to which it is attached. If the wire is bowed it will not develop its full strength. Sight down the wires to confirm their straightness after rigging. If they are not straight, the strap that connects the clevis end to the aircrafts surface should be bent so that it will be straight (Figure 4-15). Note that this relationship may change as large rigging changes are made.

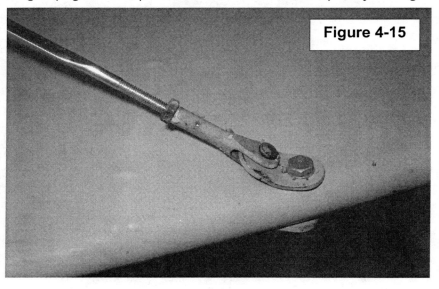

Figure 4-15

Different sizes and materials of wires have different minimum thread engagements. Larger wires appear to require that they be threaded in to a minimum of two times the diameter of the thread (Figure 4-16), however the smaller wires may only require an engagement of one diameter (1A instead of 2A in Figure 4-16). If there are witness holes, go by those to determine the minimum engagement of thread.

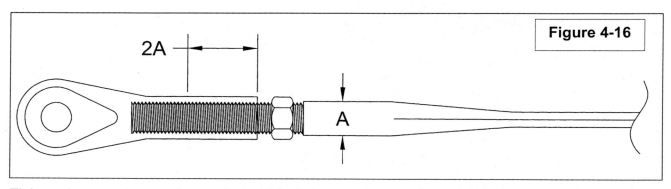

Figure 4-16

Flying wires made up of stranded cable (wire rope) are adjusted in a manner similar to solid wires but they use turnbuckles like the kind found on control systems. To prevent the mass of the turnbuckle from contributing to vibration of the wire, the turnbuckle should be located at the end of the wire rather than the middle The same turnbuckle rules given previously apply when they are used as flying wires. When stranded wire is used in parallel pairs as a support, it is essential that the tension between the pairs be the same so that they each carry an equal part of the load.

Check or Jam Nuts

The thin jam/check nuts used to secure rod ends and clevis forks (shown loose in Figure 4-16) should be torqued to the values provided in FAA <u>Advisory Circular 43.13-1B Aircraft Inspection and Repair</u>, given for shear type nuts.

Stall Strips

Stall strips are commonly used to tailor the aircraft response to a stall (Figure 4-17 and 4-18). Typically installed on the leading edge inboard, they may be employed to force the wing root to completely stall before the wing tip, allowing some roll control during the stall and causing the aircraft to begin to pitch down on its own. They are sometimes arranged asymmetrically between the two wings to make an airplane stall symmetrically, where previously it had a tendency to always roll (start to spin) in one direction. They work by forcing airflow separation prematurely, and have the beneficial effect of providing aerodynamic buffeting that warns of impending stall. The drawback of stall strips is a small loss of lift.

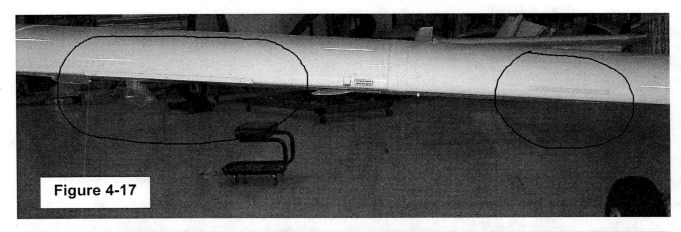

Figure 4-17

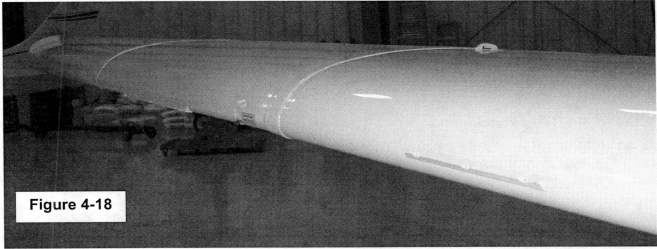

Figure 4-18

Because normal landings are made at stall speed in a light plane, it is desired that the aircraft has no tendency to depart in roll when the stall occurs, even when one foot above the runway. Stall strips are useful for making the landings easier in such planes that tend to roll rapidly after the stall.

Stall strips are triangular, typically 60° in cross-section and between a 1/4" and 5/8" on the flats (Figure 4-19). They are usually less than 12" long (they are seen smaller and larger depending on the aircraft). In general, small wing leading edge radii require strips that are small in cross-section, and vice-versa. The position of the stall strips requires experimentation and are best held on with tape until something satisfactory is found (duct tape won't stick for long in the cold).

Initially make them short and work towards longer strips, testing the stall characteristics each time. Airplanes that employ stall strips generally have one or two strips on each wing. Where there are two strips, they are typically separated by a distance of at least the length of the longest stall strip. It is preferable to start with one strip per wing. Try about 5%-10% wingspan outboard of the fuselage to start, with the strip on the most forward part of the leading edge radius (Figure 4-19).

The Grumman AA-5B in Figure 4-17 and 4-18 has a sophisticated arrangement. The shorter stall strip located inboard on the wing is set much higher on the leading edge than the longer stall strip located at midspan.

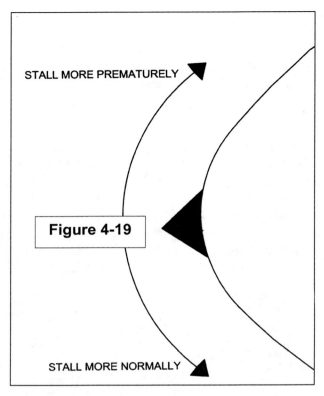

Move the strips spanwise on the wing and up or down on the leading edge to change the flight characteristics. Moving the strip up on the leading edge will trip the air sooner, making the area behind the strip stall sooner than the other parts of the wing. Make adjustments in increments of about 1/8" at a time. Some aerobatic airplanes use them further outboard on the wings to change the stall/snap roll characteristics.

Improper installation or sizing of stall strips can degrade the maximum lift capability, resulting in long takeoff rolls, poor climb rate, etc.. It is preferable to begin experimentation with a strip that is smaller in cross-section than one which is larger. Stall strips may alter the spin characteristics of the aircraft.

One can observe many examples of stall strips at the local airport. They can be fabricated from aluminum wood, glass/foam, etc…

Fixed Trim Tabs

Fixed trim tabs are usually employed for the roll and yaw axes, but may also be used to provide elevator trim. An example is illustrated in Figure 1-7 in Chapter 1. They only provide trim for one airspeed, power setting, angle of attack, and configuration. Where they are used on the roll axis, the reason is often due to improper rigging or a structural defect. On the yaw axis, they may help relieve the rudder forces due to propeller phenomenon. Elevator trim requirements vary greatly with airspeed changes and most airplanes use a cockpit adjustable trim tab because of this.

A tab on a control surface reduces the effectiveness of that control surface. It also increases the control force in one direction and decreases it in the other. The effectiveness of tabs has been shown to vary considerably between power on and power off if the tab is in the propeller slipstream. Adding tab(s) will require rebalancing of the control surface if the control surface was mass balanced (Chapter 9).

A stiff piece of aluminum (approximately .032" thick) will make a suitable fixed trim tab for most small airplanes. It may need to be thicker on a very fast aircraft. If one is finding it difficult to obtain the proper setting, or it seems like it changes slowly over time, make sure that the tab is mounted securely (the structure it is attached to is important) and that it isn't deflecting under a load.

Many examples of the sizes and shapes used can be seen at the airport. For aerodynamic reasons it is preferred that it be of a larger area with a small deflection, rather than a smaller area with a large deflection. The higher the aspect ratio, the more efficient it will be (long and thin rather than short and fat). A high aspect ratio tab will require slightly less deflection to produce the same force than a low aspect ratio tab, given that the two tabs are of equal area. However, there is some minimum chord length necessary to get the tab into undisturbed air at the trailing edge of the control. The chord of a tab should be between 5% and 20% of the chord of the surface to which it is attached (Figure 4-20). On some aircraft, the turbulence near the trailing edge of the control may be significant due to surface roughness or wing design and the chord of the tab may have to be lengthened to get it into the airflow.

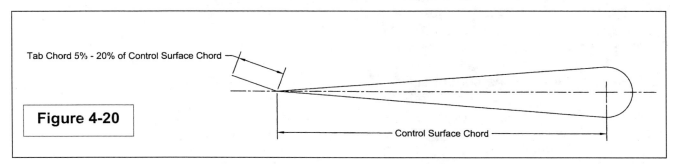

Experiments have shown that a satisfactory maximum for tab deflection is between 15° and 20°, measured against the chord line of the surface to which it is attached (Figure 4-21). If a control requires more force than that provided by the 15° to 20° deflection, it is better to increase the area of the tab. A control force reversal may occur if the trim tab is bent more than about 75% of the maximum deflection of the surface to which it is attaches, for example an elevator with a deflection of thirty degrees should have a maximum tab angle of about 22°.

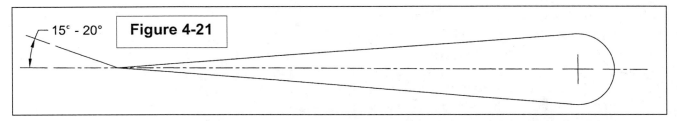

A tab used on one aileron will typically be more effective (require less deflection) if bent in one direction rather than the other. Which direction will require some experimentation. Aesthetics will favor a downward bend. Note that the effectiveness of a trim tab decreases as the deflection of the control surface to which it is attached increases.

The spanwise location of a tab along a surface is also important. In chapter 2 the spanwise lift distribution is illustrated on a wing, and is similar to the empennage surfaces as well. Because more lift is produced inboard, the aileron trim tabs may be more effective if placed as far inboard as possible. If they are affected by the propeller slipstream however, it may result in adverse aileron forces during certain phases of flight. In the case of the elevator, the trim tab may be partly in the wake of turbulent air shed from the local structure if it is too close to the fuselage. Rudder trim tabs are most commonly found in the clean air at the bottom of the rudder.

Other options that produce small amounts of trim are illustrated in Figure 4-22. All of these are designed to produce a force that deflects the control surface down relative to the figure. The bottom one is a piece of rope, relatively small in diameter, that is doped to a fabric covered surface with some fabric doped on top of that to form a 'dam' for the air.

Because the lower four methods in Figure 4-22 are relatively small in cross-section, they are designed to go over a large percentage of the span of the control surface. A side effect of these fixes is that they alter the shape and size of the trailing edge of the control surface over the whole length, changing both the feel and amount of lift produced with control deflection.

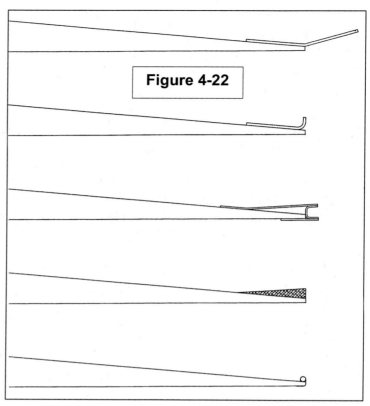

Figure 4-22

It can be a positive or negative effect. They may make a surface more conducive to flutter.

Unless some tailoring is needed to the response of a control, or it is an absolute historical reproduction, trim tabs are preferable to wedges and strips.

Some method is desired to make accurate and consistent bends on long tabs. To avoid scratching paint and twisting the tab, sandwich the whole tab between two pieces of stiff wood and clamp them in place. The tab may be then evenly bent.

Spades

Setting up and adjusting spades is as much of a design issue as rigging issue, and is particular to an individual aircraft. As such, no information is provided here but the reader is referred to the four volumes of technical tips sold by the International Aerobatic Club. The October '87 issue of Sport Aviation has a short article in the Craftsman's Corner by Ben Owen that provides information to get started with sizing and placement.

Vortex Generators

Like spades, installing vortex generators is a design issue and isn't discussed here. The reader is referred to the NASA/NACA Technical Report Server for research in this area (see References about this), as well as to the manufacturers of aftermarket vortex generators. The installation of vortex generators can have large positive and negative effects on flight characteristics.

Type Certificate Data Sheets

TCDS's are legal documents, issued by the FAA for certificated aircraft, which describe the aircraft for the purposes of certification. Included in a TCDS is the major rigging information for that aircraft. They can be downloaded by make and model from the FAA's website.

Setting Up

High wing airplanes will require a lot of work to rig if there is only one stepladder (Figure 4-23). It is well worth setting up two or more stepladders or scaffolding.

Figure 4-23

Chapter 5
Initial Rigging

Setting the Fixed Surfaces

Some tasks require that more than one axis of a surface be set simultaneously with another axis of that surface, so read through the section before beginning and make a plan. Where these procedures are used to drill the final holes for attachment of flying surfaces, a bolt should be inserted in each hole as it is drilled, to prevent misalignment.

The rigging of biplane wings is discussed in Chapter 11 but much of the necessary information is also given here.

Leveling the Fuselage

Level the fuselage laterally and longitudinally. The reference datums on the fuselage, from which the longitudinal and lateral level measurements are taken, should be evaluated for their ability to provide consistent measurements and for their access as each part of the aircraft is assembled/rigged. It may be worthwhile during construction of the fuselage to add a small permanent member(s) on which a level may be placed for the purposes of rigging. This is frequently done on certificated aircraft and information about the leveling points is provided in the maintenance data.

In the case of a fuselage being rigged to accept the main flying surfaces during construction, it should be restrained solidly. This is particularly important where no wing adjustments are possible after the spar fittings are drilled, or the wing is glassed in. It is easiest to 'cradle' the fuselage on the shop floor in a jig, without the gear attached. This allows the wing(s) to be more easily restrained on all of the necessary axes prior to permanent attachment.

The wing attach points are the reference point for leveling the fuselage laterally because the wings have the largest influence on the aircraft. Often the wing attach points are impossible to access after the wing is attached. Sometimes adapters (1-2-3 blocks) are used to offset the level or other rigging tool to an accessible location if the distance is short (Figure 5-1). Obviously the adapters must be exactly the same size.

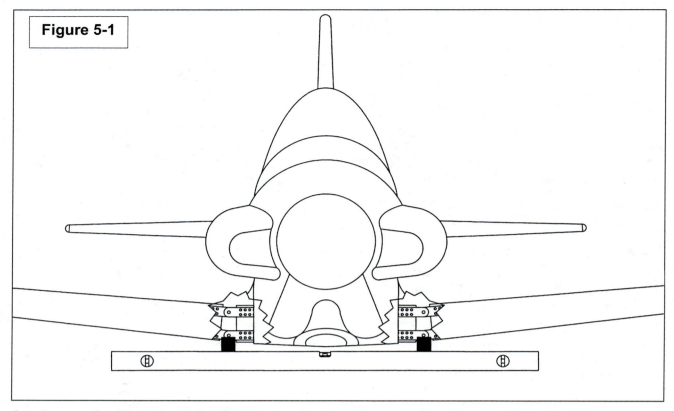

Figure 5-1

Another method is to level the fuselage using the wing attach points, and then find another part of the fuselage which provides a consistent indication of level (a perfectly straight and flat structural member, for example). It is common to add a member for the purposes of rigging.

In aircraft with a tall cabin, a common method is to drop a plumb bob from the ceiling to a pre-surveyed point on the floor. This has the advantage of leveling on both axes simultaneously, but is typically not as consistent as a level in making very accurate measurements.

However the reference datum is established, it must provide consistent measurements. It is important that it be a part of the basic aircraft structure that is not removable. It is especially important to have a consistent method of leveling the fuselage longitudinally. The incidence and thrust line measurements are based on the fuselage longitudinal axis. Consistency is more important than accuracy here because the fuselage position in relation to everything else has less effect on flying characteristics and performance than does differences between the flying surfaces and their relation to the thrust line.

The longitudinal axis of the fuselage is usually parallel to a straight section on the fuselage structure (longeron perhaps). An extra member may be added for this. Some aircraft provide for specific leveling points as in Figure 5-2 where two long bolts are screwed into preset nutplates in the side of the fuselage and a level is laid on the two round shanks of the bolts.

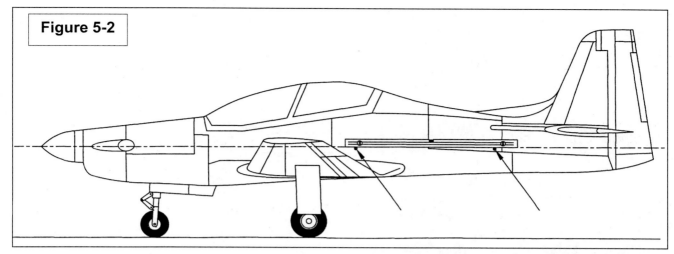

Figure 5-2

It is difficult to keep a fuselage from shifting while doing all this work, especially if it's on the gear with inflated tires. Jacks should be used to get the fuselage level so it won't move. A tail stand (Figure 5-3) is a great assistance in leveling. For a quick rigging check, it is sometimes enough to level the aircraft by deflating one or the other tire.

Figure 5-3

Setting the Vertical Stabilizer

The vertical stabilizer is an integral part of most airframes, fixed during construction and generally not provided with a means of adjustment on any axis. Some aircraft designs specify that the vertical stabilizer have an incidence angle relative to the aircraft centerline (see Chapter 1 and 2).

The vertical stabilizer is generally attached to the fuselage while the fuselage is in the assembly jig (cradle, table, cage, etc., whatever was used to build the fuselage). The fuselage should be leveled laterally and longitudinally using the guidelines provided previously, and solidly secured.

The vertical stabilizer needs to be oriented spanwise parallel to the vertical axis of the aircraft (Figure 5-4), and chordwise parallel to the longitudinal axis (or with an incidence angle) (Figure 5-5), simultaneously.

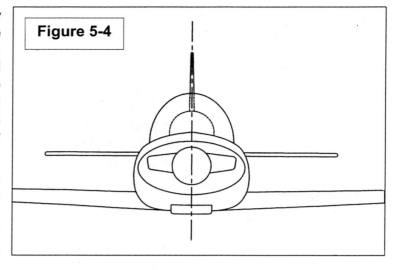

Figure 5-4

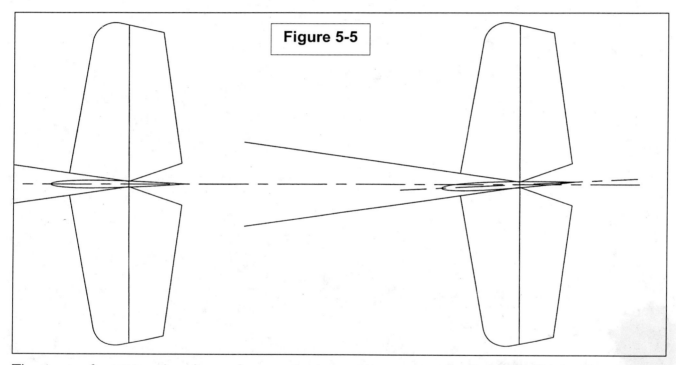

Figure 5-5

The type of construction (truss frame, semi-monocoque, composite) will dictate the order of assembly. Some aircraft will have the vertical-stabilizer-aft-spar (where the rudder goes) attached to the fuselage before creating the rest of the vertical stabilizer, while others will have a complete vertical stabilizer assembled and ready to attach to the fuselage.

The spanwise axis is easily set with a level oriented vertically on the side of the vertical stabilizer (a tapered adapter for the level may be necessary if the thickness of the vertical stabilizer decreases with span).

Where the vertical stabilizer is assembled separate from the fuselage (on a jig), the stabilizer spars are restrained in the proper orientation to each other, eliminating any twist. Where the vertical stabilizer is assembled on the fuselage, measurement of the chordwise axis (twist) requires a more elaborate setup. Observe the setup in Figure 5-6. The orientation is checked in several places along the span to ensure that the vertical stabilizer has no twist (it may be at

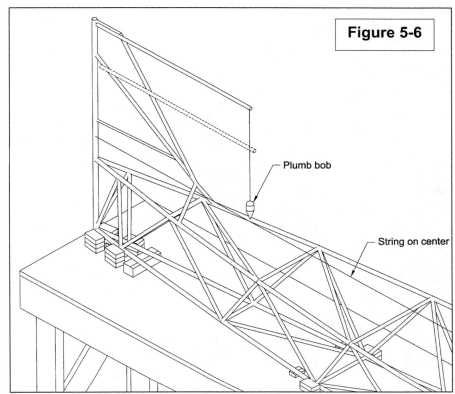

Figure 5-6

an angle to the centerline if so designed, but usually that angle is constant along the span of the vertical stabilizer).

Setting the Wing

Recheck the fuselage for level, and recheck it frequently while working. The wing is the most powerful influence on the aircraft since it reacts the most aerodynamic force. The wing adjustments must be accomplished as accurately as possible to maximize aircraft performance. In addition, other flying surfaces may be set in reference to some axis of the wing geometry.

How one proceeds is different for a two piece wing, one piece wing, high wing, low wing, glassed in wing, or bolted on wing. If the aircraft is still under construction, all adjustments are available (as long as wing spar fittings and such haven't been drilled yet for their attaching bolts or pins). It is suggested that in the absence of very precise jigs, fixtures, and fabrication methods, that the wing not be glassed in or have its attachment holes drilled until the wing is (mainly) complete and the main fuselage structure is finished. On many wings the geometry about all axes must be set simultaneously because of the method by which the wing(s) are attached to the fuselage and lack of adjustment after they are attached, so it is important to develop a plan for setting the wings.

Because of the difficulty in setting a wing about all three axes simultaneously during construction, it may be necessary to have enough tooling to measure each aspect of the geometry simultaneously, in addition to restraining the wing(s) in the correct position(s) for attachment.

The Wing to the Longitudinal Axis

The wing planform must be symmetrical about the aircraft centerline (as viewed from the top) to be most efficient. This adjustment is typically only available during construction of the aircraft.

It should be set so that a reference point close to the wing tips (precisely located points symmetrical about the centerline, typically on the last rib where the spar is) is equidistant from the aircraft centerline (at some point on the centerline as far away from the wing as possible, like the hinge line of the vertical stabilizer). See Figure 5-7, when A = B then a = b. This works on the assumption that the points being measured from on the wing, are exactly symmetrically deposed about the aircraft centerline. This method tends to average out any fabrication variation in the wings, especially if checked against another set of points on the wing further inboard. These same points on the wing will be used for making other measurements as well. This axis is easily measured with a trammel bar or tape measure.

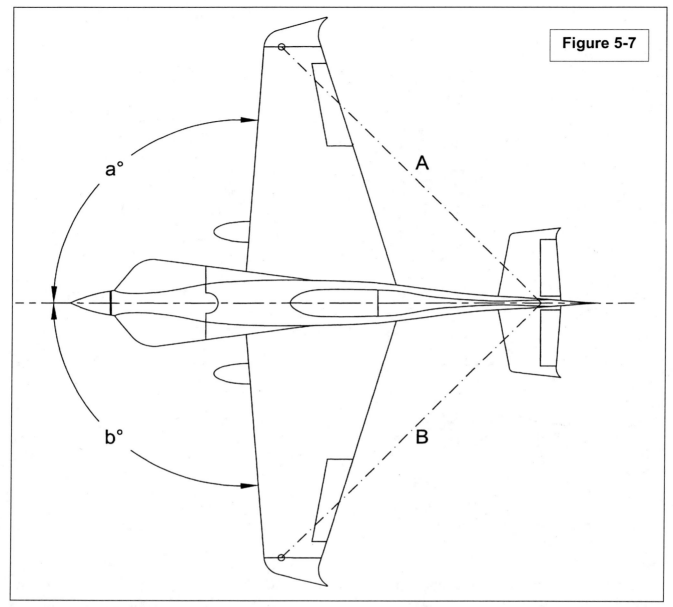

Figure 5-7

The Wing to the Vertical Axis

The next step is to set the wing perpendicular to the vertical axis of the aircraft for a wing without dihedral, or, if the wings have dihedral, dihedral will be also be set (see Chapter 1). On aircraft without wing struts, dihedral is not adjustable and is set during construction. Many high wing aircraft have threaded attach points (forks) on the struts to shorten or lengthen them, thus changing the dihedral angle.

For a low wing aircraft without dihedral, level the wing (or wings) using a level placed spanwise on the wing/wing panels. If the wing tapers in thickness from root to tip, it may be necessary to use a dihedral board even if there is no specified dihedral (see Chapter 1 and Chapter 3). The main wing spar is the preferred spot for placing a spanwise level (on fabric wings with raised rib-stitching, use some shims to hold the level/dihedral board above the stitching). Check both sides of a one piece wing with the level and divide the difference if there is any, being sure to use mirror image locations about the centerline of the aircraft for placement of the level.

For dihedraled wings, the procedure is different for a strut-braced high wing as opposed to a low or mid wing.

On a low-wing aircraft (non-adjustable dihedral), block the wings at the correct dihedral (using the dihedral board from Chapter 3) to hold it in position for setting the incidence. Note that changes in incidence (later on in this section) may affect the dihedral depending on how the wing is restrained at this step.

NOTE

If the bolt holes for the wing attach points are already drilled and found to be not satisfactory, repair or replacement may be in order.

Many high wing aircraft have a wing strut that may be shortened or lengthened to produce the correct dihedral. The procedure is somewhat different depending on whether there are one wing strut or two wing struts to a side (Figure 5-8). The addition of jury struts will affect the process of rigging the wings.

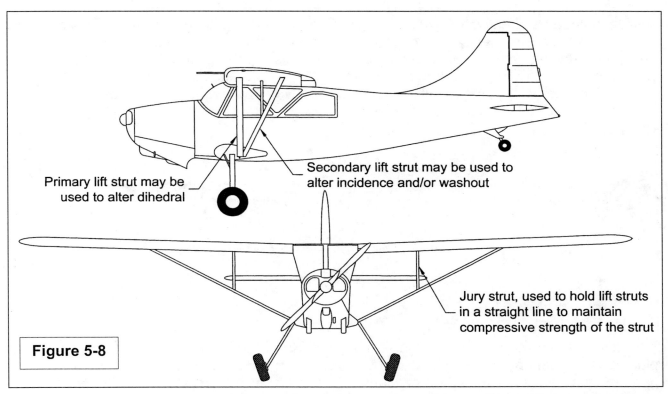

Figure 5-8

To set the dihedral on a high-wing aircraft, disconnect the secondary lift strut (if installed) and the jury strut (if installed) and loosen the bolts at the wing attach fittings. Lengthen or shorten the primary lift struts to produce the correct dihedral. It may be necessary to put the weight of the wing on a stand while setting the dihedral (where jury struts are used). Otherwise, the primary lift strut will bow out under the weight of the wing, changing the dihedral and incidence later when it is straightened with the jury strut. When the dihedral is correct, attach the forward jury strut to the primary lift strut and adjust it so the primary lift strut makes a perfectly straight line (not bowed or curved under the weight of the wing). If the lift strut had a significant bow or curve before straightening, the dihedral will change (increase) upon straightening the lift strut.

Some designers/manufacturers specify that a string pulled between the upper wingtips will have a certain vertical distance from the wing attach points when the dihedral is correct (Figure 5-9).

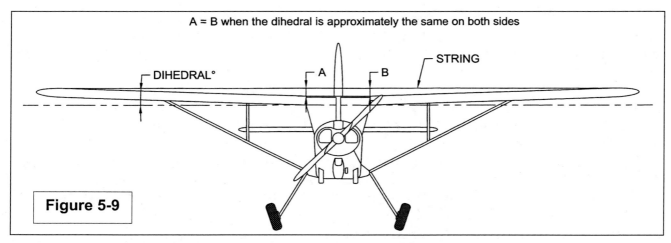

Figure 5-9

This technique has relatively low accuracy and can be time consuming to adjust since moving one wing up or down affects the string for both wings. For many small aircraft flight envelopes, small inaccuracies here may not produce a noticeable difference in aircraft flying qualities or performance. Refer to the example in Figure 5-10. A difference in string height of only 0.09" between the two wing roots resulted in approximately 3/4" difference in height at the wing tips and 0.4° difference in dihedral.

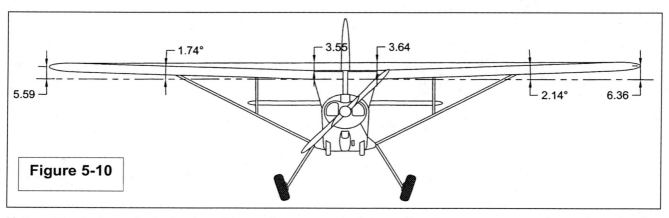

Figure 5-10

If the dihedral angle is known (it can be converted from the linear measurement given in the plans using trigonometry), a dihedral board like the one discussed in Chapter 3 will provide greater consistency between the two wings and make it much faster to set the dihedral.

Wing Incidence Angle

The designer will specify an incidence angle. The wing incidence is normally exactly the same on both sides of the aircraft. Wing incidence (in relation to tail incidence, thrust line, and fuselage longitudinal axis) has a significant impact on flying qualities and performance (Chapter 2). Any difference in wing incidence between one side of the aircraft and the other will seriously degrade performance and reduce control authority. Because of fabrication tolerances in the wings it is important to make several incidence measurements on each wing at different spanwise locations, before deciding on the necessary adjustments to match the two wings.

On many aircraft the incidence is not adjustable and the wings must be assembled to the fuselage with the correct incidence. This can be very difficult to do accurately, considering the need for the wing to also be correct on the other axes simultaneously (lined up with the spar fittings and with the correct dihedral), and the need to actually drill and bolt the wing attach points while the fuselage and wing are set in their proper relation to each other. A few aircraft have an adjustment at the aft spar fitting to allow the changing of incidence (a bolt to turn or eccentric pin to rotate). Some wings may be adjusted for incidence by the use of shims. See Chapter 3 about tools for measuring wing incidence.

On wings with washout, the designer will have specified a certain wing station at which to measure the incidence since the incidence angle changes along the span of the wing (reduces towards the tip). Washout or no washout, it is most accurate to compare the incidence at several wing stations while setting the wings and divide any discernable difference (consider that the actual difference is not purely geometric because the amount of lift produced by the wing is at a maximum at the root and tapers off to nothing at the tip with a fairly elliptical distribution, see Chapter 2).

Washout

On most aircraft washout is fixed in the construction of the wing and is not adjustable. Some high wing aircraft have a secondary lift strut (Figure 5-8), making washout adjustable by lengthening or shortening the secondary lift strut (the wing is fairly flexible in torsion on these designs). Biplanes may also use the interplane strut for adjusting washout (see Chapter 11 about biplanes). After the dihedral is set and the forward jury strut attached (if installed), use an appropriate incidence board at the specified spanwise location to set the washout or incidence. Reattach the jury strut to the secondary lift strut and check the secondary lift strut for straightness. Adjust as necessary and recheck the washout since removing the bow from the secondary lift strut is going to change the washout.

Setting the Thrustline

Unless something different is specified in the plans (or called for in the course of experimentation with the rig), the thrustline should initially be set parallel to the longitudinal axis of the aircraft (see Chapter 1). Level the aircraft laterally and longitudinally in the same manner that was used to set the wing. Because of flexibility in the vibration dampeners of the engine, the engine may sag slightly. Excessive sag is caused by worn dampeners or a poor mounting system.

NOTE

The process by which an engine is mounted will affect the thrust line. See Chapter 7.

See Chapter 3 on tools for measuring the thrustline. Small corrections to the thrustline may be made by inserting a steel disc or shim of the desired thickness between the engine and rubber mount (Figures 5-11, 5-12, and 5-13), or between the firewall and the engine mount (least desirable). The disc must be at least the same diameter as the flanges on the engine mounting pads or the mount will not be developing its' full strength. The shims must be making contact all around their circumference. If the thrustline is found to be grossly off in all directions, it would be safer to fabricate a new engine mount. Tightening the mounting bolts for the engine mount onto adjusting shims of greatly varying thicknesses may preload the engine mount and/or vibration dampeners excessively with stress, and change the frequency response of the vibration dampeners.

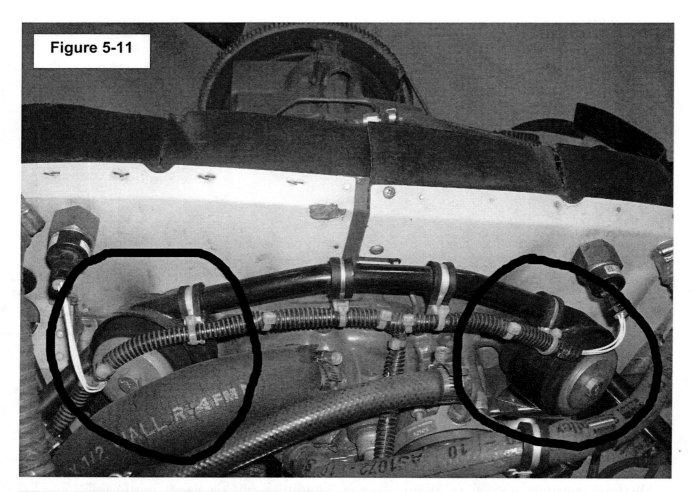

Figure 5-11

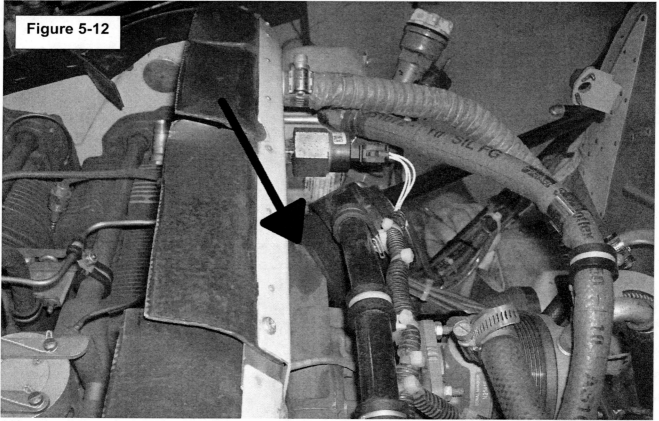

Figure 5-12

Figure 5-13

Vertical Thrustline

The thrustline may be checked vertically in several ways. The most accurate methods are described in Chapter 3. As an alternative, a level may be placed on a part of the engine case known to be parallel to the crankshaft. If the propeller has a fairly flat working surface on the front side, a level may be oriented vertically on the propeller (propeller vertical) in a manner similar to the method given in Chapter 3 with the test bar. Levels are less accurate in the vertical position.

Horizontal Thrustline

Like setting the vertical thrustline, there are several methods of setting the horizontal thrustline. The most accurate methods are described in Chapter 3. In addition, sighting down the split in the crankcase should reveal any gross deviations from the aircrafts longitudinal axis.

To get a little more precise, fabricate a bracket to hold the string over the crankcase splitline at the front of the engine, such that the string is as close to the engine as possible. Pull the string to the rear of the fuselage and fasten it on the centerline of the fuselage. Look straight down on the string to see if it is lined up with the crankcase splitline. Sometimes the crankcase splitline is only parallel to the crankshaft and not exactly on its center.

Setting the Horizontal Stabilizer

The horizontal stabilizer is set much in the same way as the wing.

Check the distance from the wingtip or aircraft centerline, to the tip of the horizontal stabilizer on the same side with a trammel bar, as was done for the wing in Figure 5-7. It should be the same as the opposite side.

With the fuselage leveled from the same points that were used for the wing, set the horizontal stabilizer parallel to the wing or wing axis using a level set spanwise on the horizontal stabilizer. If the horizontal stabilizer is not flat on top (tapered in thickness), a dihedral board will be necessary with the level.

Incidence angle will be set with an incidence board, or, if the longitudinal axis of the airplane coincides with the upper longerons, the angle may be converted to a linear distance and measured up from the longerons. Like the wing, checking incidence in several spanwise locations and dividing the difference will produce the best results.

If the stabilizers are braced with wires, the vertical and horizontal stabilizers are accomplished simultaneously. See the paragraph on adjusting flying wires.

Setting the Slip/Skid Indicator

The inclinometer is installed in reference to the wing. Level the aircraft in the same way that was used to set the wing, and install the inclinometer so that the ball is exactly in the middle. It will be most accurate and stable if located on the aircraft centerline (buttock line 0). See also Chapter 3.

Setting the Movable Surfaces

Aileron and Elevator

Before beginning, back off all the control stops at the control surfaces and controls to allow the maximum mechanical movement. If there are multiple adjustments in the control system, (threaded rod ends for example), set all of the adjustments at the midpoint of their allowable travel. While working, take up adjustments a little from each available point in the circuit, rather than all the adjustment at one point.

Adjusting the controls is typically a three step process;

1) set the neutral positions of the control surfaces and control actuators

2) set the cable tensions (if cables are installed)

3) set the travel.

Steps one and two are frequently performed simultaneously.

The first step is to set the control surfaces and control actuators at neutral. The control surfaces are clamped in their neutral position (with neutral boards or whatever, see Chapter 3), and then the pushrods and/or turnbuckles are used to center the stick/wheel in the middle of its travel. Another simple tool for holding a control surface at neutral is illustrated in Figure 5-14. Where cables are involved, turnbuckles on both cables are necessary to set the position of the stick/wheel prior to cable tensioning.

Figure 5-14

Setting the neutral positions may actually be done with the controls fixed or the control surfaces fixed, depending on the aircraft or whatever is easiest. Sometimes it is easier to lock the control actuators (stick, yoke, pedals) in their neutral position and adjust to bring the control surfaces to neutral. However, when done in this order and where cables are involved, the weight of the control surface on the cable will create some deviations in the finished rig. In a two cable closed loop system, locking both the actuator and the control surface and then adjusting equal cable tension is also a method used, but may result in slight deviations of the neutral position due to small inaccuracies in setting cable tensions.

Where a bellcrank is used as the interface between cables and a pushrod, the bellcrank must also be adjusted to the center of travel around its radius by adjusting the length of the pushrod, prior to adjusting the cables for centering the stick/wheel (Figure 5-15). Sometimes it is easier to lock the bellcrank and adjust both the control surface and actuator to neutral.

NOTE

The elevator control actuator is not usually set at the center of travel as most elevators have different upwards and downwards deflections. The position of the stick/wheel is specified in the instructions or determined experimentally; where plenty of extra movement is available for the stick/wheel, in relation to the allowable elevator movement, the stick/wheel is set for the most comfortable position in relation to the pilot seating.

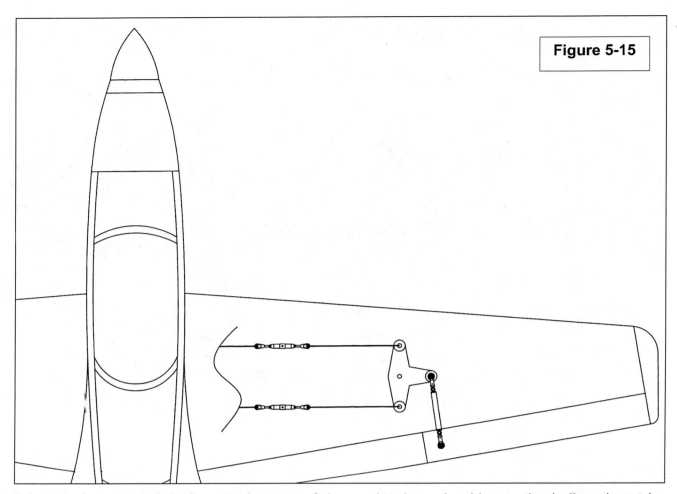

Figure 5-15

It is very important that the attachments of the pushrods and cables to the bellcrank not be allowed to go over center of the bellcrank rotational axis (Figure 5-16) during their full travel as the controls may become locked. It anything is close to going over center, airloads may drive the control surface and bellcrank further than desired, especially in the absence of control stops. In addition, as the bellcrank horns get closer to going over center, less leverage on the controls is available to the pilot and heavy airloads may make it impossible to return the controls in the opposite direction, even when the bellcrank horns haven't yet gone over center.

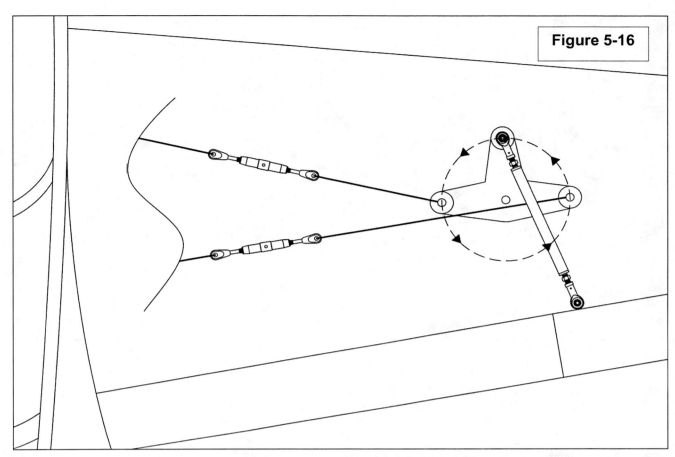

Figure 5-16

Cable tensioning is usually accomplished without the weight of the control surface interfering. On a straight cable system (without pushrods or bellcranks), the control surface remains clamped or supported. In the case of a bellcrank with cables and a pushrod attached, the control surface may be disconnected from the pushrod to remove the weight, or the bellcrank locked. In either case, the cables are then tightened to the appropriate tension, a little bit at a time on each cable to prevent the stick/wheel from being altered from the neutral position (alternatively, locking both the bellcrank and actuator and adjusting both cables to equal tension may be acceptable)

When the controls have been rigged to neutral positions and the cables tensioned, the control travel is set by the stops *at the control surface*. Then the stops at the actuator (stick or pedals) are adjusted so they contact slightly after the control surface stops. Powerplant and secondary flight control stops are also rigged in this manner. The reasons for this are explained in Chapter 4.

It may be wise to wait until all controls have been rigged and checked before applying safety wire and similar stuff.

Setting the Aileron Travel

The neutral position of aileron trailing edges are almost always set inline with the wing trailing edge, or, so that the wing airfoil is shaped as designed (see airfoil geometry in Chapter 1).

With the ailerons at neutral, set the protractor head on one aileron and adjust it so the bubble is in the middle (Figure 5-17).

Figure 5-17

Remove the protractor head and note the number of degrees for neutral. Adjust the scale for the desired aileron deflection (many aircraft have different up and down deflections for the ailerons). Replace the protractor head and move the aileron until the bubble is centered (Figure 5-18).

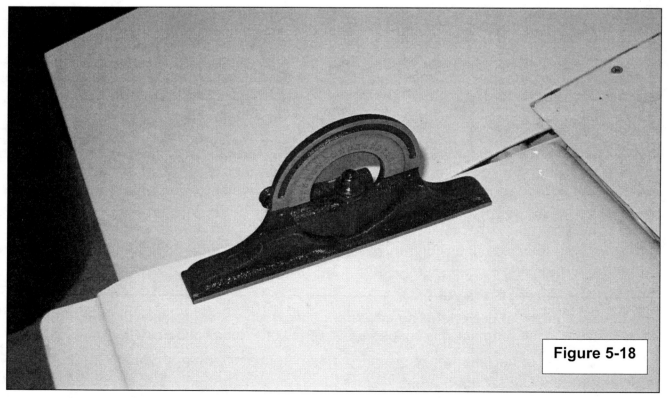

Figure 5-18

Adjust the control stop so the aileron cannot move any further in this direction. Reset the desired angular deflection on the protractor for the opposite deflection and repeat the process to set the stops for the opposite deflection on the same aileron. Always put the protractor head back in the same place on the aileron that was used for setting the 0° position. The other aileron should have about the same deflection when it is checked, and its' stops are adjusted in the same way (both ailerons should hit the stops approximately simultaneously). If there is a gross difference between the ailerons, one should investigate why since the setting of one causes the other to be essentially fixed (it is a function of the mixing ratio set by the bellcrank assembly geometry, although loose cables or a bellcrank which wasn't exactly centered can cause some difference between ailerons). If the difference is tolerable, divide the error between the two sides. Changing downward travel on one side changes the upward travel on the other.

Once the stops at the ailerons are set, the aileron-axis-stops at the stick/wheel should be set so they contact slightly after the aileron-stops. How much of a gap is left at the stops at the stick is determined by the type of control system (Chapter 4).

Setting the Elevator or Stabilator Travel

The elevator is set in the same way as the ailerons; adjust it for neutral, and then set the control stops for the correct travel. Like horizontal stabilizer incidence, the allowable elevator travel may be critical to safety, whether it moves too much or too little. The deflection of servo or anti-servo tabs should be confirmed (both the direction and magnitude). If the servo/anti-servo tab doubles as a trim tab, it will probably be rigged after the elevator or stabilator.

Elevator Trim

When the elevator is done, set the trim tab to geometric neutral and adjust the actuator to the center of its' range. Make a temporary mark to indicate neutral. Flight testing will produce the proper place for the neutral mark. On a nosewheel airplane, it is desired that there be some aft stick pressure required for the rotation and lift-off. A properly trimmed tailwheel airplane during takeoff will raise the tail slightly and then fly off the ground, without elevator input (some airplanes are different). That behavior changes with CG position in either type of airplane. The elevator trim tab is usually adjusted after the elevator is rigged but while the elevator is still clamped in the neutral position. It is common for elevator trim tabs to have different upwards and downwards deflections.

Rudder

A protractor head cannot be used on the rudder, so some other method must be devised to check the travel. A throwboard may be fabricated as in Chapter 3. A protractor may be used if it can span the chord of both the rudder and vertical stabilizer (and have an appropriate surface to work from). Alternatively, a tripod on the floor with a pointer placed near the trailing edge of the rudder in the neutral position will provide a reference from which linear measurements may be made. The rudder travel angle may have to be converted to a linear distance using some math to calculate the chord of an arc (Appendix A). See Chapter 1 on linear measurements of the deflections of control surfaces.

The procedure to rig the rudder to neutral varies from aircraft to aircraft depending on whether it is an open circuit or closed circuit control system. Some instructions may also specify that the rudder be rigged some number of degrees right or left while the controls are neutral (Chapter 2).

If there is a steerable nosewheel or tailwheel involved, it is sometimes easier (or necessary) to disconnect it until rudder rigging is complete. Landing gear rigging is discussed in Chapter 10 and usually occurs after rigging of the flight controls.

Open Circuit System

An open circuit system has one cable going from each rudder pedal to the appropriate horn on the rudder, and the two systems are independent, that is they are not connected to each other except at the rudder. Springs are used to pull the pedals toward the firewall and tension the cables.

For an open circuit system, three rigging factors must be considered simultaneously, and the adjustment of one affects the others to a varying degree. The three factors are spring tension, pedals in proper position, and rudder in proper position.

Lock the pedals in their neutral position. To set them to neutral, clamp a straight board across their faces so that they are flush with each other, then arrange a second clamp to hold both pedals at their center of travel (or whatever is neutral, it may be slightly forward or aft of their center of travel). Adjust the turnbuckles on the cables to center the rudder with very little tension. Release the clamps on the pedals and adjust the spring tension to force the pedals to their neutral position if they are not already. Spring tension is usually adjustable by providing several mounting holes for the spring at the rudder pedal (Figure 5-19).

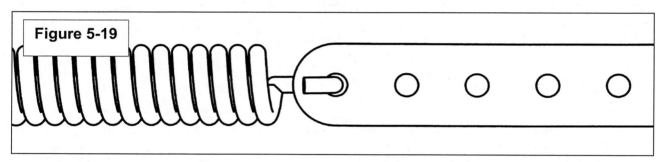

Figure 5-19

Closed Circuit System

A closed circuit system has the two pedals interconnected with a third cable and the tension is adjusted with a turnbuckle. There may or may not be springs to center the rudder.

For a closed circuit system, disconnect the short cable connecting the two pedals. If there are springs, disconnect them. Lock the pedals in their neutral position. To set them to neutral, clamp a straight board across their faces so that they are flush with each other, then arrange a second clamp to hold both pedals at their center of travel (or whatever is neutral, it may be slightly forward or aft of their center of travel). Adjust the turnbuckles on the cables to center the rudder with very little tension. Release the clamps on the pedals and attach the short cable connecting the rudder pedals. Adjust the cable tension for the whole circuit using primarily the turnbuckle on the short interconnect cable. Reattach the spring(s) and adjust their tension to center the pedals and rudder. Spring tension is usually adjustable by providing several mounting holes for the spring at the rudder pedal (Figure 5-19).

Adjusting Travel

Set the throwboard or protractor in the appropriate location, or for a linear measurement, set the tripod so the pointer is as close to the trailing edge as possible, with the rudder in the neutral position. As an alternative to using a tripod with a pointer, a piece of wire taped to the elevator and bent around to the rudder trailing edge to indicate neutral will be sufficient. Deflect the rudder in one direction until it is at the correct angle for maximum deflection (see calculating chord of an arc) and adjust the stop. Repeat in the other direction.

Set the stops now at the rudder pedals so that the rudder stops contact slightly before the pedal stops, with normal force.

Flaps

The flap positions influence an aircrafts behavior and flight performance to a large degree if their neutral positions (flaps up position) are not the same on either side. There are many different flap designs so it is impossible to provide specific rigging instructions. Some flaps have considerable span along the wing and the neutral position should be checked in several places on each flap to minimize their influence on the aircraft when they are retracted. The flaps should also be checked for symmetry at the various deflections so they don't cause a rolling moment when deflected, reducing control authority. Flaps may be checked with a neutral board (Chapter 3) or a protractor when the aircraft is leveled. It is often necessary to need to make small flap adjustments after an aircraft is first flown, especially on aircraft with large flaps.

Considerable care should be observed in flap rigging to ensure that no errors will result in jamming or breakage, that will result in asymmetrical flap deployment. A large difference between two flaps will easily exceed the rolling moments able to be produced by the ailerons.

Setting the Other Trims

There are many different trim systems and it is up to the rigger to determine the best method for rigging in the absence of documentation. Usually the primary control is rigged first, then the trim for that control circuit is rigged. Most cockpit adjustable trim tabs change geometry (deflection) as their primary control surface moves, so the position of the primary control is generally specified to be neutral when setting the trim limits. Trim tabs are generally not effective at deflections greater than 25° (Chapter 4).

A force trim system commonly used on rudders may be interrelated to the nose-wheel or tail-wheel steering and must be considered during rigging.

Ground adjustable trim tabs should initially be set to their geometric neutral (streamlined position).

Bungee and Spring Centering Systems for Primary Flight Controls

As discussed in Chapter 2, because of the variety of applications no guidelines are provided here and the manufacturers/designers documentation should be consulted. The correct tension is important as it affects the basic controllability of the aircraft in certain situations.

Aileron/Rudder Interconnect

This is used on some aircraft, usually fast cross-country cruisers or transport type aircraft that are to be flown IFR with a two-axis autopilot. These systems rely on extra cables, in conjunction with springs or bungees, and must be rigged in accordance with the designers instructions.

Wheel Fairings

Wheel fairings on small aircraft can create an asymmetrical aerodynamic force. Some method of achieving symmetry between the two fairings is desired. They should be aligned with the relative wind at the desired flight envelope. Even when they are symmetrically applied about the aircraft centerline, they will contribute to both lift and drag in various flight conditions and some experimentation with the exact orientation may be worthwhile.

Wing Tips

The aerodynamically shaped non-structural wing tips (Figure 5-20) on most aircraft are large enough to create moments that require trim to compensate. There are a wide variety of shapes and some imagination will be necessary to enable them to be rigged symmetrically.

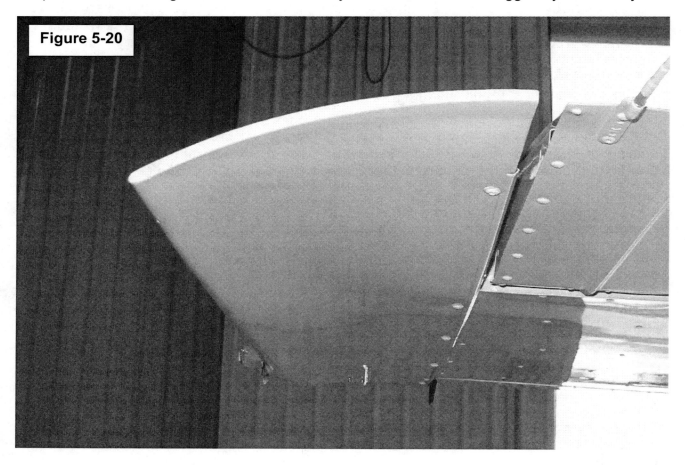

Figure 5-20

Setting Wire Tensions

In the example in Figure 5-21, an empennage is braced with wires, a common design feature of small aircraft.

The main objective of wire adjustment is to obtain proper wire tension, but in the process to establish perpendicularity between **e** and **f**, and maintain a straight hinge axis along **e** and **f** so that no binding in the hinges of the elevator or rudder occur because of a warped structure (the rudder and elevator should be attached for tensioning so they can be checked for binding). The vertical stabilizer, line **f**, should be plumb with the fuselage leveled.

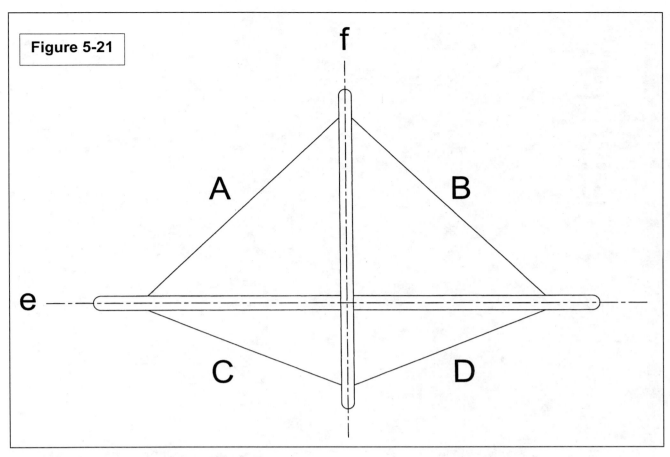

Figure 5-21

The general procedure is as follows;

Level the aircraft laterally. Tension all of the wires so that they are equal, not particularly concerned with geometry just yet, but being careful not to seriously distort anything in the process. Then working on one side of the aircraft only (in this example the left side of the picture), adjust the lower and upper wires (**A** and **C**) so that the horizontal stabilizer (**e**) on the left side is level, and all of the wires have equal tension (note that one wire must be loosened to allow the opposite wire to be tightened). Repeat for horizontal stabilizer on the other side.

When the horizontal stabilizer is set, rig the vertical stabilizer. Using a level oriented vertically (on one side of **f**), adjust the wires **A** and **B** until **f** is plumb.

The reasoning for the above order is that the lower attach points of the lower wires are usually the fuselage structure, which won't distort like the more flexible stabilizers during tensioning of the wires. It is the 'origin' of wire tensioning.

For aircraft with two sets of wires as illustrated in Figure 5-22, do the above procedure on the forward set of wires first (with the level placed near the leading edges), then do the aft set of wires with the level placed near the trailing edges. In placing a level on fabric covered surfaces, be careful of the location such that surface irregularities like stitching don't interfere with the level.

Figure 5-22

Landing and Taxi Lights

Landing lights should be oriented so they illuminate the *projected impact point* during the final approach. The projected impact point would be the point of impact if no roundout for landing was performed. The normal visual cues for initiating the roundout occur in the area that the ground is first seen *clearly*, the projected impact point.

The angle that the light should make with the longitudinal axis of the aircraft depends on the approach attitude of the aircraft. On a normal approach, a conventional airplane will typically be pointed in its' direction of travel when the flaps are deployed. A conventional airplane on a short-field approach may be nose-up in relation to its direction of travel. Airplanes designed for STOL operations may be either way depending on the design and the landing light should be illuminating the point of touchdown for those aircraft, since such landings occur without much rollout. This may cause the landing light to be projected upwards during rollout and taxi, eliminating forward vision. Either some compromise must be made with the one light, or it may require two lights (landing and taxi).In any case, it will probably take some experimentation to determine the correct angle for a particular airplane.

Taxi lights should be angled to provide a comfortable view when taxiing.

Compass

The compass should be lined up with the aircraft. If it's designed to be installed in an instrument panel, it's assumed that the panel is perpendicular to the aircrafts longitudinal axis.

For a non-panel mount, it should be laid out so that it is oriented properly with the longitudinal axis of the aircraft. Although some adjustment is available with compensating magnets, it is relatively little and isn't enough to compensate for large installation errors.

The procedures for swinging a compass are given in aircraft mechanics textbooks.

Testing the Rig

Once the aircraft is rigged and everything is safetied, fly the aircraft and make sure it is controllable. After experimenting with the rig, it would be best to approach the flight envelope limits in an orderly fashion before attempting any extreme maneuvers. Chapter 2 discusses some of the important things to test for.

Before experimenting with the rig for the purposes of increasing performance, make a note of the power setting required for the desired flight envelope. Make a note of the control positions and/or forces required to operate within the desired flight envelope. Improvements may be quantified in this manner and negative changes will be noticed before they become dangerous. Textbooks on flight test engineering provide methods of comparing performance when measurements are taken under different atmospheric conditions.

Chapter 6
Correcting Rigging Problems

Rigging problems usually manifest themselves as roll or yaw, or both (sideslip). Other rigging problems are usually of a more serious nature and Chapter 2 should be consulted as a place to start.

The difficulty with identifying rigging problems is that a change in rotation or translation about one axis of an aircraft usually results in one or more changes about other axes. This requires a systematic approach to isolating problems.

If an aircraft that was previously rigged well seems to suddenly have a rigging problem, find the reason before continuing. Most rigging issues involve the primary structure and flight controls, and either a sudden or slow change in the rig is indicative of a serious problem.

Without knowledge of the aircrafts prior rigging condition, making changes to compensate for a rigging problem may degrade the control authority and reduce the performance of the aircraft. It's not hard to get an airplane to fly straight and level, but getting it to point in the same direction that it's traveling with the controls approximately neutral requires more work. It is necessary to check the complete rig of an unknown airplane, perhaps resetting it to the initial recommended condition before diagnosing rigging problems.

Because of the wide variety of airplane configurations and designs, it is only possible to make generalized statements about what to do in a particular situation. Many experimental aircraft do not display the flight characteristics of certificated aircraft. A good knowledge of aerodynamics and flight mechanics are necessary to diagnose many apparent rigging problems.

Almost all rigging problems vary with speed and power and this fact can be used to advantage in determining the problem. The following paragraphs provide several ways of looking at the same problems. When testing the rig, it is helpful to write down what is going on with the airplane for easier recall later, even when the problem seems simple.

Straight and Level Flight

An aircraft will diverge eventually at some higher speed if there is even a tiny asymmetry, so limit the tests to the speed ranges which are important. Initially set the aileron and rudder trim at geometric neutral (if installed). The wing fuel should be evenly distributed. However, in a small side-by-side seated aircraft with one pilot, it may be desired to distribute the fuel load to even out the weight. See Chapter 2 about the effects of offset lateral CG.

It may be helpful to test the rig initially with power-off, in order to establish a baseline for separating propeller phenomenon from aerodynamic misrig. A rigging problem that occurs only with power-on, regardless of speed, is from adverse propeller forces. By diving the airplane to some speed with power-off, a comparison may be made to the power-on condition at that same speed. A maladjustment of a flying surface that lies in the propeller slipstream will be greatly magnified at slow speed with power-on, and should also appear at high-speed with power-off.

If the aircraft tends to roll hands off, especially with increasing airspeed, the problem is probably wing incidence, wing washout, or a misadjusted flap; anything that changes the lift on one wing. Any of these situations will create yaw in the opposite direction of roll, and fixing the roll problem will often make the yaw go away.

Although many airplanes will roll due to yaw, the rudder deflection and subsequent angle of yaw required for this to happen is fairly large and the rate of roll is slower than if the rolling moment is caused by the wing. Some experimental airplanes will roll due to yaw only when the yaw rate is great.

When yaw occurs because of roll, the yaw is generally in the opposite direction of the roll. When roll occurs because of yaw, roll and yaw occur in the same direction.

If the airplane rolls and yaws in the same direction, an adjustment is probably required on both the roll and yaw axis. In this case it will probably take several iterations of adjustment to arrive at the final solution since roll affects yaw and vice-versa. A vertical stabilizer set with incidence will always cause yaw to occur at some higher speed.

If some angle of bank is required to maintain a constant heading then the aircraft is slipping and the ball will be out of center (if the instrument is set properly). The aircraft has stabilized in a slight slip, and roll and yaw are opposite each other. This may be a problem on the yaw axis. Push on the rudder to center the ball and try it again. If the aircraft then stabilizes on a constant heading with the wings level and the ball in the middle, the problem probably lies mainly on the yaw axis.

An aircraft which changes heading very slowly with the wings level and ailerons neutral may be a yaw problem, and the ball will be deflected in the opposite direction of the turn. Hold the ball in the center with the rudder pedal and note the behavior of the aircraft. If it does not roll and continues in straight flight, the problem is on the yaw axis. Not all aircraft roll with yaw.

The yaw string provides the best indication of the translation of the aircraft, but finding a suitable spot for it is difficult on a single-engine tractor airplane.

Moving the ailerons with a trim tab will cause a yaw away from the direction of roll. This may be needed, or it may just make a new problem. Adjusting a flap will probably also cause yaw, but will make more change in rolling moment than yawing moment, compared to making a change in aileron deflection. Changing wing incidence will cause a lot of roll and yaw.

The objective is not just to make the airplane fly straight, but to make it fly straight with the controls approximately neutral, and aligned with the relative wind. The propeller slipstream may prevent this but some compromise will be found through experimentation to achieve the proper rig in the desired flight envelope.

A common indication of flying in a slight slip is that the fuel drains from one wing tank into the other over a period of time. Some slips are barley perceptible.

Ailerons will align themselves with the local relative wind and will not center exactly in a stable slip or roll if left to themselves, unless the two ailerons are very different from each other in the neutral position. When there is friction in the controls one can stabilize in a slip or roll even when things are rigged properly.

Bendable Trim Tabs

Installing bendable trim tabs like the ones discussed in Chapter 1 and Chapter 4 should be done when it is determined that no rigging problems exist. The control actuator will be off center when the control is left to the influence of the trim tab(s), but an attempt to recenter the controls with linkage adjustments may cause problems with control authority in other flight regimes.

NOTE

Adjusting the neutral position of a single aileron in an attempt to trim in roll isn't effective, the ailerons simply seek equilibrium with the stick off to the side, and no rolling moment is produced.

When adjusting bendable trim tabs, it's a good idea to keep track of its position with a protractor, so it can be performed in a methodical manner.

Stalls

An aircraft which tends to roll in one direction only during a stall can be challenging during takeoff, landing, and spins. Because of the effects of the propeller, it is only possible to investigate aerodynamic symmetry with the power at idle. Because an asymmetric lift condition will be also greatly reflected at high speeds, problems with asymmetric stalls may be traced to something that is causing airflow separation, or the propeller slipstream. An airplane which always rolls in the same direction during power-off stalls and can't be controlled easily may have problems getting out of spins in that direction.

A flap which is deployed for the stall may be the cause of the lift difference between the two sides, even though it isn't reflected in high speed flight as a rolling moment. This occurs because of certain flap linkage geometries which are non-linear in their action; a small difference between the two flaps in the neutral position may be magnified when the flaps are extended.

To accurately test stall characteristics, an airplane should be decreased in speed (using pitch) at a rate no faster than 2 Knots per second. This makes it easier to compensate for pilot error. Some aircraft have a sharp drop in lift at the stall, which magnifies small errors during stall entry. On these aircraft it may not be a concern to the pilot if it goes off in one direction all the time, if it is controllable and the airplane has no other objectionable characteristics.

A *sometimes* effective way of determining the aerodynamic symmetry is to do a falling leaf; keep the aircraft in a stalled state (holding the stick back) while using the rudder to keep the wings level (defying it too spin). If the airplane remains controllable its probably fine.

Determine how rapidly it recovers from right and left incipient spins, power off, a 1/4 or 1/2 turn at first. Proceed with the spins with caution if the asymmetry hasn't been fixed.

NOTE

Aircraft which have the vertical stabilizer rigged at an angle may have a noticeable difference in yaw power between the right and left and power-on and power off.

If the reason for the airflow separation is not obvious, it is usually easier to force airflow separation sooner on the other wing. See Stall strips in Chapter 4.

In some cases it may be found that an aircraft is not able to stall or develop its' maximum lift. The rigging of the horizontal stabilizer and elevator are not providing the necessary control power to use all of the available lift of the wing. This can be caused by incorrect rigging of the horizontal stabilizer incidence or elevator deflection. Refer to Chapter 2. It is also possible that the CG is too far forward.

Maneuvering Flight

Problems in maneuvering flight are difficult to separate from the propeller forces. To separate propeller influence, climb to a high altitude and do some maneuvering with the power off at the intended rigging speed, trading potential energy for kinetic energy. Some airplanes with large powerplants will still feel slightly asymmetric at idle or zero thrust. Airflow separation rarely occurs perfectly symmetrically, and induced drag rises very rapidly with increases in angle of attack, so trying to split hairs when accelerating at high angles of attack is going to be frustrating. If though it is found that much aileron/rudder deflection is required with power-off at high angles of attack to maintain equilibrium, then a rigging problem probably exists.

Distortion of a somewhat elastic structure caused by pulling g's may significantly change the behavior of the aircraft and can be confused with a rigging problem.

High Speed Flight

Compensations taken in the rig that provide for control in low-speed/high-power flight, are frequently seen in the dive. One can only compromise with the low-speed rigging to provide satisfactory high speed characteristics.

A rolling moment at high speed that isn't obvious at low speed is probably incidence, washout, or flap, although aerodynamically shaped wingtips that are slightly asymmetric may be the culprit (at low speed with power off this may be reflected more as a yawing moment than a rolling moment because of the rapid rise in induced drag).

Distortion of the flying surfaces due to high dynamic pressure can result in large trim changes or odd characteristics. A structure which lacks rigidity (although still being strong enough), is going to distort under aerodynamic forces, changing its position and hence the aerodynamic forces that it produces (known as aeroelastic divergence). Fabric covered control surfaces are most subject to this and it manifests itself in several ways.

The distortion of the fabric can cause an increase in camber of the flying surface, magnifying lift or changing its direction. This is usually caused by maladjustment of the incidence of the horizontal stabilizer, or, the intentional setting of the vertical stabilizer at an incidence to the aircraft centerline. Because the offset of the vertical stabilizer or a certain incidence may be necessary on a particular aircraft, that distortion under high dynamic pressure may be one of the limiting factors related to V_{NE} (never-exceed speed).

Another common distortion is the bulging out of the control surfaces equally on both sides, caused by the airflow impinging on an opening and pressurizing the inside of the control surface. This can significantly change the amount of control authority available, and cause violent up and down oscillations of the control surface due to aerodynamic overbalancing. This may also lead to flutter (flutter generally occurs at a higher frequency than overbalance oscillations).

Chapter 7
Vibration

The elimination of as much vibration as possible can have a positive effect on aircraft performance and increase the life of the aircraft components. Consider the following factors;

1) Powerplant vibration requires energy, reducing the thrust horsepower of the engine and propeller combination.

2) Vibration transmitted to the airframe surface may induce airflow separation, reducing lift and increasing drag. A vibrating propeller and/or vehicle is continuously attempting to shed the air that is flowing around it.

3) Vibration in one component may instigate or sustain vibration in another component.

4) Vibration may cause the engine to alternate position constantly, continuously changing the direction of the thrust line.

6) Aerodynamic vibration (separating and reattaching airflow) is associated with increased drag and possibly reduced control effectiveness when the control surfaces are in the wake of the unsteady airflow.

7) Vibration wears out mechanical parts and materials (fatigue life).

Vibration on an aircraft is usually caused by the engine, propeller, or combination of both. Vibration may also be caused by airflow, noise, or another vibration.

Sudden changes in vibration level are an important cue that something about the aircraft has changed and it should be identified (doing an engine run-up with a tailwind component sometimes suddenly produces a noticeable vibration).

Elements of Vibration

Vibration is an oscillatory motion. It is an unbalanced force, or system of forces acting on or through an elastic or resilient material or structure. All objects or systems of objects that are elastic and have mass are capable of vibration. Almost all of the materials used in aircraft construction fit this description. This includes items like crankshafts, which aren't normally thought of as bending during operation.

The oscillatory motion of vibration is *periodic*, it repeats itself in equal time intervals. Most vibrations are *harmonic*, that is the graph of the vibrations smoothly transition from one point to another (Figure 7-1), and are easily described mathematically by the circular functions, sine and cosine. The graph in Figure 7-1 is a sine wave. Harmonic motion occurs frequently in nature, whether describing the frequency of a particular color of light, the action of a spring, or the pulsing of a star.

A vibration is described in terms of its amplitude, cycle or period, and its frequency (Figure 7-1).

Amplitude
The amplitude is the amount of physical displacement from the resting position of the vibrating component(s).

Period and Cycle
The period is the *time* interval in which the motion repeats itself. The cycle is the *motion* that completes itself in one period.

Frequency

The frequency is a measurement of the number of cycles/periods that occur in a certain amount of time. Mathematically, the frequency is equal to the reciprocal of the period. Common units for frequency are;

- Hz - Hertz, a frequency of 100 Hz means 100 cycles per second.
- CPM - Cycles Per Minute
- CPS - Cycles Per Second
- RPM – Revolutions Per Minute

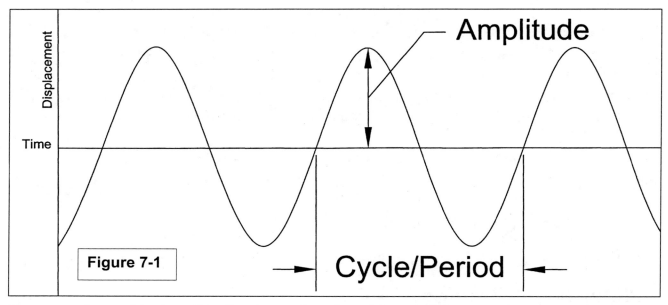

Figure 7-1

Frequency is used to describe vibration, rather than the period, because it more accurately reflects those things which are being described, like the rotational speed of the engine for example.

Classes of Vibration

Vibration is generally classified in two ways, free vibrations and forced vibrations.

A free vibration occurs when an excitation to an object is applied and removed (a quick tap on a tuning fork for example), and the vibration eventually dies out. Figure 7-2 illustrates an example of a free vibration.

A forced vibration is one where some excitation is sustained (a running engine for example) and the vibration remains continuous. Figure 7-1 is an example of a forced vibration.

Excitation is the term used to describe the force or forces that cause vibration. If the force varies with time, it causes sustained vibration of the objects that are subject to the force. The rate at which the excitation occurs is known as the forcing frequency, and the objects that it influences will vibrate at the forcing frequency.

When discussing mechanical vibrations, the excitation is assumed to be external to the object that is vibrating (an engine causing vibration of aluminum skin for example). This is unlike aerodynamic flutter, which is a self-exciting vibration (Chapter 9). Excitation may come from the propeller, engine, noise, or aerodynamic phenomenon.

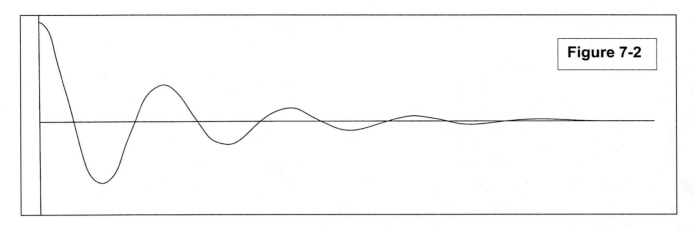

Figure 7-2

Natural Frequency

The natural frequency of an object (or system of interconnected objects) is the frequency at which it will vibrate in free vibration. The excitation is assumed to be instantaneous and abrupt, like the tap on the tuning fork. The natural frequency of any particular object is determined by its' material, stiffness, and mass/mass distribution. Mathematically, the natural frequency is a function of the static deflection of the object (static deflection being that displacement from the resting position due to the objects own mass). As a general rule, things which are relatively flexible tend to have low natural frequencies while something stiff will have higher natural frequencies, this is fairly intuitive. Less intuitive is the relationship between the frequency and amplitude. For the relatively small amplitudes found in harmonic vibration, the amplitude is not affected or related to the frequency. For a given object, as the amplitude increases, so does the restoring force that accelerates the object back in the other direction, making the period of the cycle remain constant for different amplitudes. In forced vibration, an object will vibrate at the frequency of the forcing excitation and not the natural frequency of the object. However when the forcing frequency and natural frequency are equal, the amplitudes of the object rise very rapidly, a condition termed *resonance* (discussed later).

To complicate matters, there may be a number of natural frequencies for a particular object (but usually only the lowest frequency is of importance). The higher natural frequencies above the first one are known as *harmonics* and are generally multiples of the natural frequency (sometimes fractions of the natural frequency, subharmonics, are important). Figures 7-3, 7-4, and 7-5 illustrate the vibration modes (shape of the bar) of a 12.0" long, 0.5" diameter steel bar at the first natural frequency and several harmonics. In the following illustrations of a *modal analysis*, the bar is fixed at the left lower end and free at the other. The computer is calculating the natural frequencies and the shape of the bar when excited at those frequencies. The displacement shown is not the actual amount of displacement, but is exaggerated somewhat to illustrate the shape.

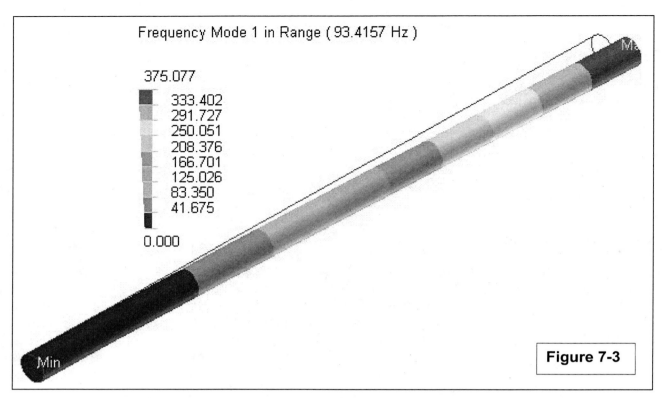

Figure 7-3

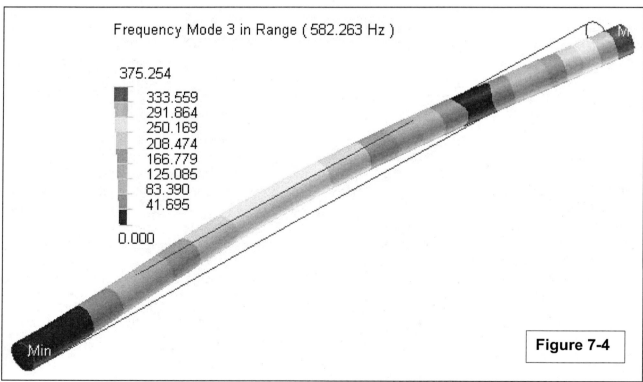

Figure 7-4

The natural frequency of an object can be altered by changing the stiffness, mass, or mass distribution. The tube in Figure 7-6 is the same length as that of the bar in Figure 7-3, and contains the same amount of the same material, hence the mass and mass distribution are the same. Because a tube is stiffer than a solid bar, it has a higher natural frequency (the other mode shapes for the tube are similar to those pictured for the solid bar except one that is unimportant for this discussion)

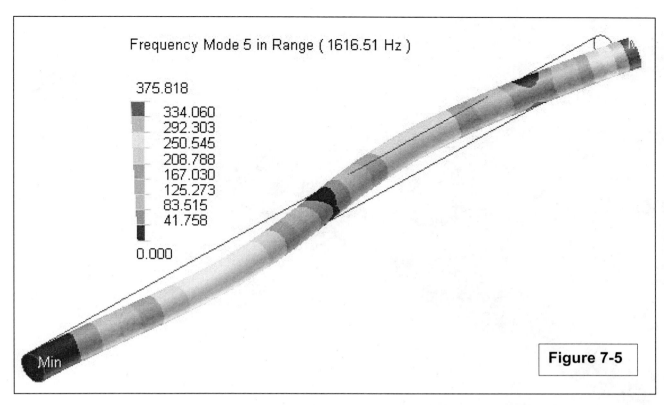

Figure 7-5

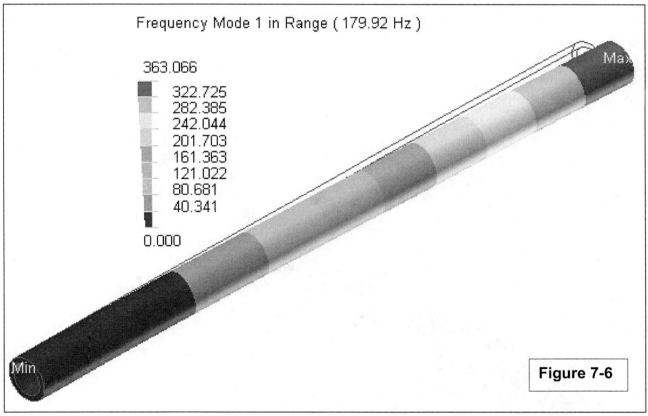

Figure 7-6

Changing the mass and distribution of mass will alter the natural frequency depending on the location of the mass. In Figures 7-7, a 2.0" diameter steel ball is affixed to the end of the tube shown in Figure 7-6, lowering the natural frequency.

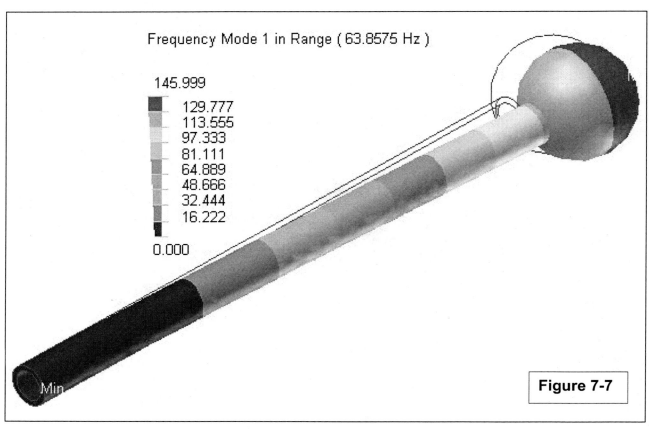

Figure 7-7

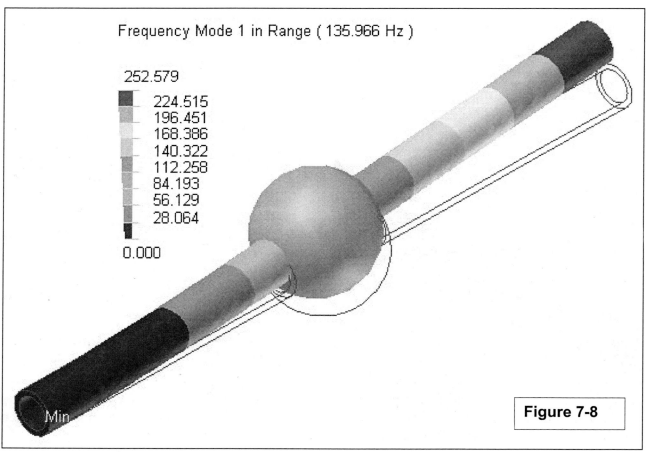

Figure 7-8

In figure 7-8, the steel ball is moved to the center of the tube, raising the natural frequency to twice that of the tube in Figure 7-7.

When an object or system has the possibility of moving or vibrating in more than one direction (has more than one degree of freedom), there will be a natural frequency for each degree of freedom. The degrees of freedom of vibration are the same as described for an aircraft in flight, three translational and three rotational (Chapter 2). Vibration in two or more degrees of freedom may be combined to create a new vibration, known as coupling.

Structures which are homogenous, like welded tubular steel or one piece composite fuselages, will have different responses to excitation than one that is bolted or riveted together from parts.

Damping and Loss Factor

Damping is any effect that tends to reduce the amplitude of oscillations of vibration. The graph of the vibration in Figure 7-2 could be an example of a damped oscillation.

Viscous damping is encountered by a body moving at some speed through a fluid, like aerodynamic flow or a hydraulic shimmy dampener. The damping increases with velocity.

Coulomb damping is caused by the sliding of dry surfaces, like the mechanical shimmy dampeners used on nosewheels and tailwheels.

Solid damping is due to internal friction within the material itself, the friction is created by the molecules rubbing each other. Solid damping of materials undergoing free vibration may be quantified as *loss factor*, the time it takes for the vibration to decay a certain amount. Wood and composites have high loss factors, that is their vibrations die out quickly, while metals have relatively low loss factors. Solid damping is proportional to the maximum stress (deflection) of the vibration cycle. The vibrations of objects which have higher natural frequencies will decay more rapidly than those with lower natural frequencies.

Artificial dampening, as is seen in engine mounts, shimmy dampeners, shock absorbers, springs, etc., is designed to reduce the amplitude to something which is considered acceptable, in the presence of continuous excitation. This is discussed in more detail later.

Resonance

When the forcing frequency (excitation) and the natural frequency of an object start to equal each other, the displacement (amplitude) and subsequent stresses on the vibrating object increase very rapidly. Imagine a forcing vibration is being used to excite a metal bar. The forcing vibration starts at a constant amplitude/low frequency and is increased in frequency, all the while maintaining a constant amplitude. In response, the metal bar vibrates at the forcing frequency, starting at a low amplitude and increasing in amplitude in response to the increasing frequency of the forcing vibration. As the forcing frequency approaches the natural frequency of the metal bar, the amplitude of the vibrating bar reaches its maximum. This occurs even though the forcing vibration has had a constant amplitude throughout the whole frequency range. A popular example is the opera singer breaking a wine glass.

Resonance occurs within a relatively small range of frequencies, such that as the forcing frequency approaches the natural frequency of the object, the amplitudes of the object are relatively small and rise slowly. Then suddenly the amplitudes increase very rapidly to a maximum, and then decrease just as rapidly as the forcing frequency increases beyond the natural frequency of the object (Figure 7-9).

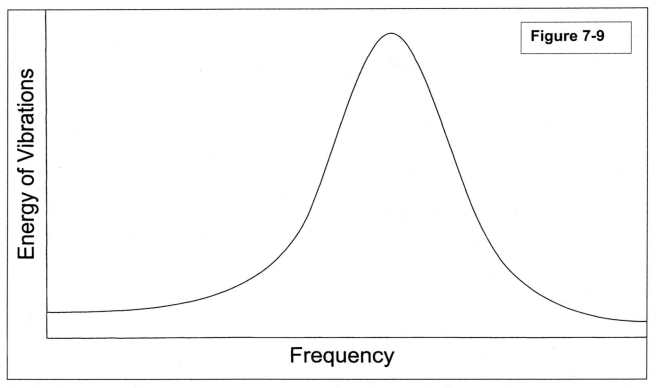

Figure 7-9

Machines are generally designed to avoid resonance. Sometimes it is necessary for a machine to transition through resonance because it couldn't be eliminated during design, however either the time will be very short or some method of damping will be used, to prevent a large build-up of forces and stresses. Many aircraft powerplants have RPM ranges that are prohibited from being used continuously, often due to resonance.

Vibration of Rotating Parts

Vibration in rotating machine parts is caused by unequal mass, or unequal mass distribution, around the rotational axis. The vibration is caused by the centrifugal forces of the unbalanced mass acting on the bearings of the rotating part. Centrifugal force increases as the square of the rotational speed. Doubling the RPM will increase the centrifugal force by a factor of four, tripling the RPM will increase the centrifugal force by a factor of nine. This fact makes troubleshooting somewhat easier because vibration will increase very rapidly with the rotational velocity.

NOTE

Vibration in rotating parts can also be caused by looseness (bad bearings maybe), even when things are well balanced.

The frequency of the vibration from rotating parts is either the same as the RPM or a multiple of it. The vibration which occurs at the same frequency as the RPM is called the first order vibration. Vibrations that occur at a frequency of twice the RPM are second order vibrations, half the RPM are called half order vibrations, and so on.

Different notation is used to describe vibration of rotating parts on aircraft. 'P' rotations describe the vibration associated with the propeller, and 'E' vibrations describe the vibration associated with the engine. For example a '1P' vibration occurs at the same frequency as the propeller RPM, a '1/2E' vibration occurs at one half of the engine RPM, a '2P' is twice the prop RPM, and so on. On direct drive engines, propeller speed is the same as engine speed and P and E are the same.

The centrifugal forces are reduced or eliminated through the process of balancing. Rotating engine parts are balanced by the manufacturer, but propellers may be balanced by the factory, repair station, or mechanic/homebuilder. There is considered two fundamental types of rotational imbalance;

- Static unbalance is caused by unequal *amounts of weight* around a rotational axis.
- Dynamic unbalance is caused by unequal *distribution of weight* around a rotational axis.

Looking at the case of static unbalance first, using a propeller as an example, the propeller in Figure 7-10 has more weight on one blade than the other (the black spot on the top blade). The center of gravity (where the axis of inertia is), is different from the axis of rotation. Centrifugal force will act on one side more than the other, which will be transmitted through the crankshaft into the engine and so on. This may be rectified by balancing the propeller about its rotation axis so that each blade is the same weight.

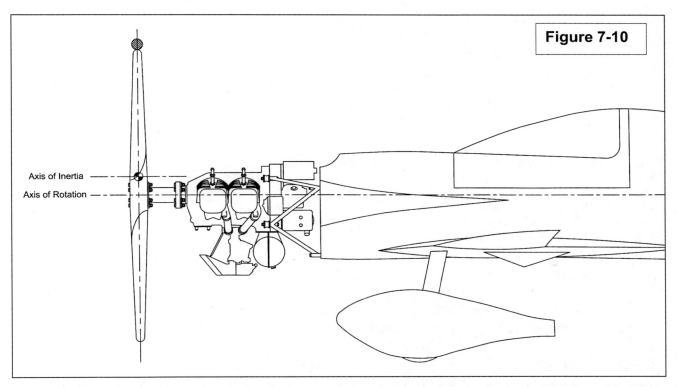

In the other case, dynamic unbalance caused by unequal *distribution of mass*, assume the weight added or subtracted from the propeller results in a little mass behind one blade, and a little mass ahead of the other blade. In Figure 7-11, the CG has moved to the axis of rotation and the prop is perfectly statically balanced, but, the mass distribution on either side of the blade has canted the axis of inertia in relation to the axis of rotation. This will result in vibration that increases with RPM (centrifugal force), even though the prop balances perfectly at rest.

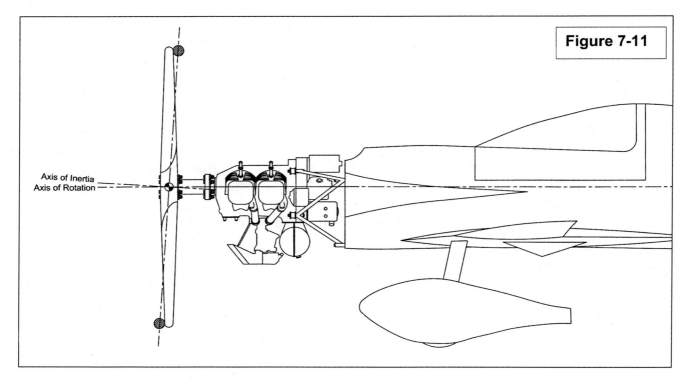

Figure 7-11

NOTE

Unequal mass distribution is typically negligible in the case of the average propeller because the very short distance from the back of the propeller to the front of the propeller presents a very small force on the crankshaft, when an imbalance exists. A slightly bent propeller will create a serious unbalanced distribution of mass (a blade bent forward or back), however the act of propeller tracking will usually identify this before dynamic balancing is performed. On something like a large belt redrive pulley, which may be eight inches from front to back, an unequal distribution of mass will impose large forces on its shaft and it should be dynamically balanced.

Static balancing is accomplished before dynamic balancing. Although something which is dynamically balanced will also be statically balanced, the equipment used for dynamic balancing probably will not be able to measure the large vibrations associated with large static imbalances and so no useful balancing information will be obtained by the dynamic balancing equipment. Dynamic balancing of propellers is discussed more in Chapter 8.

Measurement of the Amplitude of Vibration

The amplitude of a vibration may be described in several ways to fit the situation. While the end result of vibration is that the vibrating component moves some distance from the resting position, many cases of vibration do not lend themselves to being measured as a linear or angular distance. The following types of amplitude units are encountered when discussing vibration.

Displacement

An amplitude given in terms of displacement describes the actual movement of the vibrating part from its resting position. The unit is often given in MILS (a MIL is 0.001"). This type of measurement is usually used for describing vibrations of low frequencies (less than 10 Hz).

Velocity

An amplitude given in terms of velocity describes the *rate of change of displacement over time*. It is commonly stated in Inches Per Second (IPS). This type of measurement is usually used for describing vibrations at medium frequencies (10 Hz to 1000 Hz). Propeller vibration is usually measured in IPS.

Acceleration

An amplitude given in terms of acceleration describes the *rate of change of velocity over time*. This is given in units of gravity (g's). 1 g is equal to an acceleration of 32 feet per second per second. This type of measurement is usually used for describing high frequencies (more than 1000 Hz).

Relationship Between Displacement, Velocity, and Acceleration

Figure 7-12 shows the relationship between displacement, velocity, and acceleration on a constant vibration. At maximum displacement, velocity decreases to zero (the deflection of the vibrating component is reversing direction). As displacement goes through zero, acceleration is constant and velocity is at its maximum. The difference between the waves is known as the phase angle. The phase angle of velocity is 90° ahead of displacement, and the phase angle of acceleration is 180° ahead of the displacement. Understanding this relationship can help in diagnosing vibration problems.

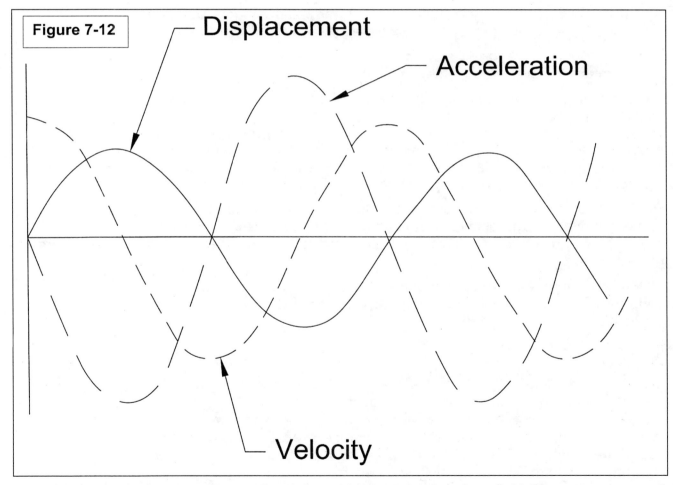

Figure 7-12

Relationship between Amplitude Measurement and Frequency

Frequency is a function of time, and velocity and acceleration are also both functions of time. Because of this, when describing amplitude in terms of velocity or acceleration, the measured amplitude will be different at different frequencies. Displacement is an absolute measurement and not a function of time, therefore it is independent of frequency.

Structural Fatigue

Structural fatigue is the progressive structural damage that occurs when a material is subject to fluctuating loads (like vibration). Those loads need not be equal to, or greater than, the strength of the material to cause failures. The fatigue life is mainly determined by the number of repetitions at a given range of stress (number of cycles at certain amplitudes). The surface finish of the piece also determines its fatigue life.

Metal propellers are the most common victims of fatigue, especially when experimenting with different prop/engine combinations, and reworking metal propellers. Propellers made out of wood or composites frequently don't have a fatigue life because of their low vibration levels.

Vibration Dampeners/Isolators

Engines, instrument panels, and some avionics are isolated from the airframe by vibration dampeners/isolators. A vibration isolator functions as a temporary energy absorber and time delay, to reduce or alter the transmission of vibration.

An isolator has two competing objectives;

1) To constrain motion: it must be strong and stiff enough to support the weight of the equipment in the presence of acceleration, and resist thrust and torque for an engine.
2) To isolate vibration: it must be somewhat flexible to absorb the vibration rather than transmit it, even when under stress from external loads.

Vibration dampeners/isolators are matched to the weight and frequency of the object they are working on. If they get worn out or if incorrect isolators are used, it can make the vibration worse than if nothing was used at all (it can also cause resonance). Because most objects have multiple isolators on the same geometric plane (four isolators on the back of an engine for example) that are selected to share the loads imposed on them, the failure of a single isolator will change the vibration response of the whole system.

Rubber is a strong material and is effective for isolating vibration, but it doesn't retain its properties over a large enough temperature range to be useful in many environments. Most isolators are now made of a synthetic that retains its properties over a wider temperature range.

The manufacturers of isolation mounts publish engineering guides on selecting the correct parts for isolating a particular vibration (see the Lord Corporation website for aircraft specific information).

Noise

Certain materials and/or structures create noise as they vibrate. Powerplants are also responsible for noise. In addition to being tiring to the pilot, noise can be a valuable cue that vibration is occurring. Noise is a vibration in the air, and can be a forcing frequency on materials or structures. An interesting phenomenon occurs when two frequencies of noise are nearly the same. The resultant noise has a 'beat'. A common example are the unsynchronized propellers of a multi-engine airplane.

Aircraft Vibration

Airframe vibration is usually caused by powerplant vibration. Airframe vibration can also be caused by aerodynamic reactions, noise, or a combination of elasticity and aerodynamics in the case of flutter.

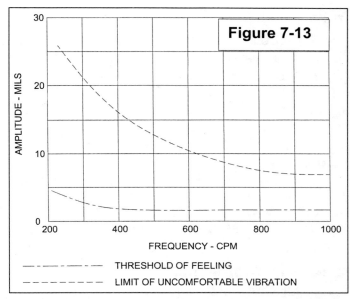

Figure 7-13

Figure 7-13 illustrates the relationship between human perception and vibration.

Sources of Aerodynamic Vibration

Vortex Shedding

Purely aerodynamic vibration is often the result of this well known phenomenon. The frequency is mathematically predictable and is a function of the shape of the object, its size, and the speed of the airflow. Figure 7-14 illustrates the vortices which are shed from a cylinder in a steady airstream. The vortices are shed alternately from either side of the cylinder. The displacement that occurs on an object when it sheds vortices in this manner is perpendicular to the airflow. Vortex shedding does not always occur, but it is a source of drag when it does occur. When considering an entire aircraft, the angle of attack (or direction of the local relative wind), and propeller slipstream will also affect vortex shedding.

Figure 7-14

Vortex shedding is commonly responsible for the vibration in flying wires. Sometimes the wires break from metal fatigue. Aligning streamlined wires with the relative wind or changing tension slightly can reduce or eliminate vibration.

When the vortices are large (shed from something big) and impinge on other parts of the aircraft, the vibration can get quite noticeable. The fixed landing gear on many aircraft may produce objectionable vibration in certain flight conditions due to vortex shedding. Fairing the gear legs will probably eliminate this.

Propeller P-Factor (Asymmetrical Thrust)

When the relative wind entering the propeller is not perpendicular to the propeller, the different propeller blades will be at different angles of attack (see Chapter 2). Each blade is producing different amounts of lift throughout its rotation, and the forces are felt as a vibration that is relative to the prop RPM, depending on the number of prop blades. This can occur in any flight condition depending on the setting of the thrustline, but for most aircraft it is prevalent at high angle of attack conditions, or when changing attitude. The propeller slipstream may excite some part of the airframe as well. As a general rule, propellers with more blades are smoother since the pulses are of a shorter duration.

Propeller Interference

Interference of the airflow at the propeller tips may cause noticeable vibration. It may be caused by the propellers proximity to the cowling or wing, or because it is a pusher propeller in the wake of the aircraft.

Certain Aerobatics and Ground Operations in Crosswinds

The vibration that occurs when the relative wind strikes the propeller disc at a steep angle (particularly from the rear) is sometimes critical to a metal propellers fatigue life. This vibration has very large amplitude and it will be obvious when it occurs. It is commonly observed on the ground when running at high power with a strong wind striking the propeller disc at an angle. Manufacturers are required to evaluate the propeller/engine combination for the destructiveness of this phenomenon.

Aircraft Skin

An aircraft skin which is vibrating for any reason is disturbing the airflow around it and decreasing the aerodynamic efficiency. Aluminum skins which are vibrating may generate noise, while a composite skin may not. Large flat areas of aluminum skin are the most subject to resonance. Usually the answer here is to either increase the stiffness of the vibrating section, or use a dampening material of some sort to change the natural frequency.

Stiffness is easily changed by adding a stiffener to the inside of the skin. Dampening materials are any of the noise reducing products available, often available in large sheets to be cut to size and glued to the inside of the skin. One common type is a foam rubber mat with foil on one side that is cut to size and glued to the inside of the skin. That all adds weight so it may be best to wait until everything else is adjusted first to see if the vibration goes away.

Cables, Hoses, Tubing, and Antennas

Cables, wiring, hoses, and tubing are highly subject to fatigue from vibration and must be restrained accordingly. AC 43.13A-1B provides guidance as to the minimum clamping that must be added. There have been a few rough running engines caused by a fuel line that was vibrating so much it caused an interruption of fuel to the engine. Engine controls, hoses, and wiring must have some short, flexible, unrestrained section between the mount and the engine, to allow for normal powerplant movement on the engine suspension system. Antennas are subject to vortex shedding and resonance and may break either the antenna or underlying structure.

Instrument Panels

Instrument panels are generally mounted on anti-vibration mounts. It is commonly thought that panels without gyros do not need anti-vibration mounts, but this largely depends on the individual aircraft. Most mechanical flight and engine instruments have very delicate springs or D'Arsonval movements that become resonant with airframe vibration, making the gauges difficult or impossible to interpret at certain engine speeds. It also wears out the instruments very rapidly.

Engine Mounts and Suspension Systems

Tubular steel engine mounts are particularly vulnerable to vibration due to their flexibility. This usually manifests itself as fatigue failure near or at the welds. It is common to see small metal straps welded between tubes to prevent cracking of the joints. These are usually an afterthought.

There are many different types and designs of suspension systems, some much better than others. They must be in good condition for them to work properly. The manufacturers of the airframe or the suspension system provide inspection criteria for determining their effectiveness. The engine must not be directly contacting the mount, as may be the case with worn isolators which allow the engine to sag and change position.

Some suspension systems are subject to installation error. If an engine is loosened slightly in its mounts to perform some sort of maintenance, the vibration isolators may gain a 'preload' during retightening, caused by the weight of the engine dragging them along an axis for which they weren't designed. That preload alters the vibration response of the suspension system. The engine should be supported at its CG with a hoist, and the aircraft level, if more than one mount is going to be loosened (Figure 7-15).

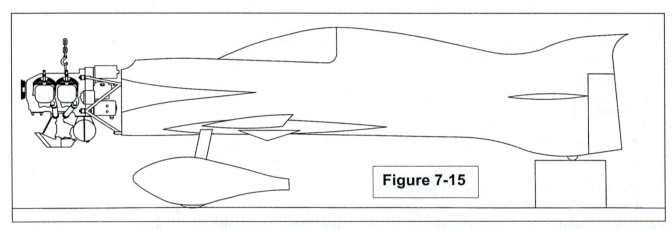

Figure 7-15

Cone shaped isolators (Figure 7-16) used on some early engine designs are the most prone to installation error. When they are installed, they should be lubricated with soapy water so they don't hang up in the engine mount bosses and get twisted out of shape while the thru-bolts are being torqued. Some cone-shaped isolators use a metal bushing inside while others don't. Those with a metal bushing inside are less conducive to installation error because the preload on the mounting bolts occurs against the metal bushing, making a more consistent frequency response among the four mounts. On those without a metal bushing, the thru-bolt has no specific preload since there is only rubber/synthetic to tighten it down on. In that case the bolts are all tightened evenly (same amount of threads showing) while the weight of the engine is taken up by the hoist. When the hoist is released, it may be seen that the weight of the engine is dragging the upper mount isolators out of their sockets on the engine case. If that happens the engine may be supported again and the bolts tightened slightly. Now there will be a difference in preload on the upper and lower isolators (whether due to bolt torquing or the weight of the engine) and some experimentation might provide a better frequency response.

On certificated aircraft that use the cone isolator mounting system with no internal bushing, a torque of between 40 and 80 IN/LBS is sometimes specified.

Powerplant Vibration

For this discussion, the powerplant is the combination of the engine and propeller. At cruise power the typical reciprocating engine vibrates at about 40 Hz (2400 RPM/60). The whole system of engine and propeller has a natural frequency of 10-15 Hz.

They are discussed separately at first.

Engine Vibration

Unequal Size and Duration of Firing Pulses

The majority of normal vibration from a reciprocating engine is caused by fluctuations of torque.

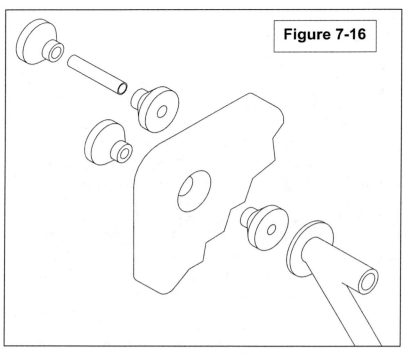

Figure 7-16

As each piston fires and pushes the crank around, the torque is momentarily increased while the piston is being driven down. The torque then decreases until the next piston gets pushed down. This is magnified greatly if one piston is not pushing as hard as the others (one cylinder not working properly). Carbureted engines don't distribute fuel as evenly as fuel-injected engines, and will vibrate more.

Unbalanced Reciprocating Components

Reciprocating components are the pistons, connecting rods, piston pins and plugs, etc.. They should be weight matched within a certain tolerance, however, they usually are done well enough from the factory. Some engines require weight-matched replacement parts.

Unbalanced Rotating Components

Unbalanced rotating components typically cause vibration at a frequency equal to their RPM (first order vibration). The crankshaft is the major rotating component and many of them have counterweights that become 'detuned' through wear. Crankshaft balancing is something that should be done by the manufacturer or a specialty shop. The starter ring gear assembly, on those engines equipped, is also balanced at the manufacturer.

Loose Components

Freeplay in bearings will result in the shaft orbiting or bending as it spins. Other loose components (including belts) may resonate at certain engine speeds.

Certain reduction drives are more prone to vibration than others. Belt drives tend to store energy, in the elasticity and free play of the belt, from either the inertia of the prop, or cylinder firing pulses. When the belt gets as much tension as it can stand, it will unleash the stored up energy. The frequency that this repeats itself on varies with the installation, or may not occur at all. All reduction drives, if some freeplay or 'backlash' exist in the mechanism, may be subject to this phenomenon. Since some backlash must exist for the proper operation of the redrive, other methods may be needed if this vibration is a problem. Certain Rotax engines use a flexible coupling between the engine and prop.

Exhaust

An exhaust which uses the aircraft structure or engine mount for support, downstream of the engine suspension system, can cause problems because the exhaust is rigidly attached to the engine and the engine is flexibly supported. A few aircraft that are designed this way have a flexible clamp (hanger) or section of the exhaust that will allow it to move somewhat independently of the engine. The flexible support of the exhaust may need many degrees of freedom to allow the engine to shift position. Exhaust systems are prone to cracking and breaking if left unsupported on the long straight sections and it is preferred to design any exhaust supports to attach to the engine rather than the mount or fuselage.

Long, unsupported straight sections of thin walled exhaust tubes are prone to vibration either from the exhaust pulses themselves, or from other excitations. Exhaust gas pulses may impinge on the airframe during certain operations, causing vibration.

Propeller Vibration

Unequal Mass Distribution About the Propeller Assembly

The unequal distribution of mass about the propeller assembly may be caused by several things;

1) The weights of the individual propeller blades vary (Figure 7-10).

2) The distribution of weight about the prop isn't coincidental to the axis of rotation (Figure 7-11). This is usually negligible if the propeller is undamaged and tracks properly. Some propellers that have been reworked may have problems with unequal mass distribution.

Aerodynamic Imbalance

Aerodynamic imbalance is normal whenever the inflow of air to the prop is at an angle to the propeller disk (P-factor in Chapter 2). Improper installation or adjustment of the individual propeller blades can cause aerodynamic imbalance to be worse, or to occur all of the time.

If the individual blades do not track on the same plane, they will produce different amounts of lift, in addition to causing the mass imbalance discussed in the previous paragraph.

If the angles of the individual blades vary, they will produce different amounts of lift. A fixed pitch prop may suffer from this as well because of damage or improper overhaul.

Resonance

Resonance may occur between the propeller and the engine, which acts on the prop or crankshaft to fatigue the metal. Certified aircraft are tested for this during development. Homebuilts may be a problem in this regard. Type Certificate Data Sheets (Chapter 4) will describe what combinations of props and engines have been tested together. Wood and composite propellers are typically not a problem because of their damping/loss factor. Resonance between a prop and an engine will occur even if the propeller is properly balanced.

There have been a number of (fatal) accidents caused by metal propellers breaking and separating from the aircraft, caused by using untested combinations of engines and propellers. They were attributed to metal fatigue caused by propeller resonance. In most of these cases, the propellers or prop/engine combination were experimental in nature and hadn't been evaluated for resonance with a vibration survey. Propeller resonance is often not detectable without specialized equipment because of its high frequency and low amplitude.

Spinner Unbalance

Spinner unbalance is caused by the same factors as other rotating components. A good spinner will be balanced by the manufacturer. While the small size of the spinner may lead one to believe that it won't significantly contribute to vibration, the difference of a few grams between one side and the other may produce very noticeable vibration. This point will be made very clear while performing dynamic balancing.

Make sure the spinner is symmetrical around its circumference, i.e., the prop cutouts are the same size. The metal should be a uniform thickness (this can be a problem on some cheap ones). Make sure the spinner backplate and front support (if installed) are centered on the crankshaft. A dial indicator can be helpful. With the spinner mounted, measure the wobble of the nose while the prop is turned. Try and adjust the spinner so there is no discernible movement between it and a fixed object in close proximity to the spinner (a camera tripod or something like it works well as the fixed object. It is frequently the case that the spinner may need to be reoriented on it's mounting to get the nose to stop wobbling.

When removing or replacing the spinner, make an index mark so it always goes back on in the same spot (this is also the case for the backplate in relationship to the prop and crankshaft). The spinner screws and their washers should be all the same size.

Propeller Flutter

Propeller flutter is typically associated with torsional vibration of a blade. Flutter phenomenon is discussed in Chapter 9. It is tested for on certificated engine/propeller combinations to ensure that it doesn't occur, but propellers which have been reworked or mounted on an untested powerplant may flutter during certain operations. Propeller flutter typically occurs during high power operation on the ground or during reverse operation on landing.

Loose Parts

Propellers with separate blades and hub are subject to wear and damage. Some manufacturers tolerate looseness of the blades in their constant-speed hubs, and the tolerance may be got from the manufacturers instructions for continued airworthiness. Ground adjustable propellers generally should have no looseness, and if there is, damage is probably already done. An overspeed can cause propeller damage that isn't obvious to a casual inspection.

Airframe/Powerplant Resonance

It is usually obvious when airframe/powerplant resonance occurs because it tends to happen in a narrow range of RPM's, and it can be felt and seen as engine shaking (on the ground). A rough test for this type of resonance is to increase the RPM's gradually from idle to full power on the ground, observing the movement of the engine. Note at what point(s), if any, the powerplant starts to vibrate or shake, and when it goes away. Repeat the test from full power to idle power, again noting the points at which vibration starts and stops. It may have to be confirmed inflight, but those RPM ranges where the engine shakes noticeably should be avoided. If it shakes all the time and gets worse with increasing RPM, something else is wrong. If the resonant RPM range is narrow and can be avoided in continuous operation, it doesn't necessarily need to be fixed.

Troubleshooting Powerplant Vibrations

It is difficult to determine whether powerplant vibrations are due to the engine or propeller. On an experimental combination the cause of vibrations may be the whole system of prop, engine, and vibration dampeners. The first thing that should be checked is the vibration dampeners between the engine and mount. If they are correct and in good shape, observe the propeller hub while the engine is running between 1,200 and 1,500 RPM; it should rotate on a horizontal plane. If the prop hub appears to swing in a slight orbit, the vibration is probably caused by the propeller. If the prop hub remains fixed then the problem is probably engine vibration.

If a propeller vibrates, whether due to balance, blade angle, or track problems, it typically vibrates throughout the entire RPM range (assuming there is no resonance between the prop and engine). The intensity of the vibration usually increases with RPM.

Vibration can sometimes be lessened by reindexing the propeller to the crankshaft. The propeller can be removed, rotated 180°, and re-installed. Homebuilders have the option of trying any propeller position, which can have a very large effect on powerplant vibration. It is suggested to put a two-bladed propeller in line with the crank pins (prop horizontal when pistons are at top and bottom of their stroke) to reduce certain types of vibration, however, experimentation may produce better results. If the engine is to be handpropped, the prop may need to be in a favorable position for that.

The propeller spinner can be a contributing factor to an out-of-balance condition. One indication of this would be a noticeable spinner "wobble" while the engine is running. This condition is normally caused by inadequate spinner front support, or a cracked or deformed spinner. Try the above test with the spinner removed. See the paragraph on spinners earlier in this chapter.

If vibrations persist, the crankshaft may be slightly bent, perhaps from a sudden stoppage. The prop flange may be measured for run-out as an initial indication of straightness, but this doesn't guarantee the crankshaft to be straight, the engine must be disassembled for the crank to be fully evaluated for straightness.

Engines whose prop hubs/prop flanges are bolted to a tapered shaft or crankshaft are subject to machining, installation, and contamination errors. This usually shows up as a tracking problem when the propeller is mounted. Often a bad fit is caused by a small burr or ding on one of the mating faces. A small ding caused by some impact will raise the metal around the depression, making a poor fit. In this case, grinding the high areas down into the shallow depression may be acceptable.

It is important to ensure that the tapered fit provides a minimum of 70% contact area, approximately symmetrically around the crankshaft, or the hub will eventually move in relation to the crankshaft. Even if the hub doesn't move, the small contact area of the hub on the crankshaft may not provide enough stiffness to prevent eventual looseness and vibration.

NOTE

Checking the contact area can be easily done, but requires some care and maybe practice to get useful results. The contact area is tested with a brightly colored, non-hardening paste-like substance. It is usually bright blue and comes in a toothpaste tube; Hi-Spot is one brand. It is carefully painted on to one part or the other with a fine artists brush, to a thickness of less than .001", and then the hub and crank are mated and torqued. The parts are then carefully separated and inspected. The transfer of Hi-Spot from one part to the other indicates where the parts are making contact, and how much. Painting the stuff on evenly and with a thin coat is the most difficult part of this procedure. If any thick spots are allowed, the stuff extrudes everywhere and gives a false indication of contact area. By saying thick spot, its meant that it is a little thicker in one spot than another, but this is still hardly measurable. It is important to mate and separate the parts very carefully to avoid transferring Hi-Spot to areas it might not otherwise have had contact (do not allow the hub to rotate separately from the crank as it is removed).

Repairing a poor fit between the hub and crank can sometimes be fixed by hand lapping the hub on, if the problem was small and mostly symmetrical. It is impossible to accurately hand lap a lot of material off and end up with the propeller flange being perpendicular to the crankshaft (meaning the propeller will track properly later). The fix is that the hub and/or crank are reground at the machine shop to a matching taper and/or the hub is machined concentric to the crank.

Chapter 8
Powerplant Rigging

The information provided in this section is general in nature and is intended to acquaint the rigger with the process of hanging the propeller and adjusting the engine. Much of the material in this section is reprinted directly from FAA handbooks and Advisory Circulars listed in the References. Chapter 7 discusses the causes and identification of vibration, and a set of aircrafts mechanics textbooks provide background information on the mechanics of powerplant systems. The manufacturers' installation and maintenance information should be consulted during the actual work.

As defined in Chapter 1, the powerplant is taken as the engine and propeller combined, and they are discussed separately in this chapter.

Propellers

A propeller is one of the most highly stressed components on an aircraft. During normal operation, 10 to 25 tons of centrifugal force pulls the blades from the hub while the blades are bending and flexing due to thrust and torque loads and engine, aerodynamic and gyroscopic vibratory loads.

Propeller rigging is a three or four-part process. It is typically accomplished in this order (depending on the manufacturers instructions, the first two steps may be switched around);

1) Adjustment of individual blades for correct angle, on adjustable pitch or constant speed propellers

2) Static balancing

3) Blade tracking

4) Dynamic balancing.

Most small aircraft only have had the first three steps accomplished, and in many cases this is sufficient. Dynamic balancing is not available at every FBO, but it is possible to rent the equipment if no places are available to perform it. Dynamic propeller balancing doesn't just balance the prop, it balances the powerplant as a whole because the prop, spinner, and engine are rigidly attached to each other.

Before starting this process, clean and inspect the propeller and spinner assembly for damage or defects. See AC 43.13-1B, and AC 20-37E about propeller maintenance.

Setting Blade Angle

Some constant-speed or variable pitch propellers must be disassembled to adjust the blade angles individually.

In setting the blade angles, consistency between the blades is more important than accuracy of the blade angle. Obviously some minimum accuracy is desired, but getting the best performance from a ground adjustable prop is usually a result of experimentation with blade angle. Therefore a measuring system and tool which has the repeatability to make all the blades the same and be able to make new measurements later based on the old measurements is the most important. For lack of detailed manufacturers instructions;

- Level the thrust line of the aircraft (make the propeller disk plumb) if the prop is installed on the aircraft. If not, restrain it solidly on a table so that the propeller disk is level.

- If the propeller is on the aircraft, rotate the blade being measured to the exact same position every time before checking blade angle. This may be done by using a level oriented spanwise on the blade.

- Clean the surface of the propeller where the measuring tool goes and remove high spots caused by dings.

- Use the measuring tool on the *exact* same spanwise blade position on every blade, making marks if necessary. The prop manufacturer will probably specify at what spanwise position the blade angle should be measured from (it is usually at the 75% span measured from the center of the prop, where the *pitch* is measured from, discussed later).

- If individual propeller blades exhibit some freeplay that is permissible, force each blade by hand to the same position when making measurements.

It will be very difficult to diagnose vibration problems, or perform dynamic balancing, if the blades are at slightly different angles.

For the purposes of rigging, blade angle is usually measured from the bottom of the airfoil of the propeller (known as the thrust surface, where relatively high pressure air is acting to produce thrust), where the surface is relatively flat. Some manufacturers will provide a sort of incidence board for this purpose, especially if the blade is not flat on one side, or the measurement is to be taken relative to the chordline or zero-lift line of the propeller airfoil.

Static Balancing

Propeller static unbalance occurs when the center of gravity of the propeller does not coincide with the axis of rotation (Chapter 7). Static balancing should occur at the propeller manufacturers before it is sent to the user. It is the process of adding or removing material from various parts of the propeller and hub so that it if it were allowed to rotate freely about its' center, it would have no tendency to do so. Essentially all the blades weigh the same. It requires having the ability to suspend the propeller assembly from its center, such that it is allowed to rotate **without any friction forces**. Since this is generally done at the factory, no static balancing is necessary on a new propeller. An older propeller may benefit, especially if it has been modified by a homebuilder or has been dressed many times to remove gouges. It is not unknown to receive a new propeller that has been incorrectly balanced.

If the protective coating used on some propellers has been repaired or replaced, it should be statically balanced (after all other work has been performed). If the propeller has been painted, it may need to be statically balanced (after all other work has been performed).

The equipment required to statically balance a propeller is not particularly sophisticated and may be fabricated or assembled inexpensively. It also may be available at the local aircraft maintenance shop. The process and description of the equipment is given here, but the removal or addition of material to a propeller assembly must be done in accordance with the propeller manufacturers repair data. Some manufacturers do not provide this information and require that the propeller be returned to the factory for the work to be performed.

Static Balancing Equipment

The method required to suspend the propeller without friction forces requires a perfectly cylindrical steel shaft through the exact center of the propeller (known as the arbor), and two flat rectangular pieces of steel (knives). The knives are known as such because one edge is sharpened or thinned to reduce rolling friction. This is illustrated in Figure 8-1. The sharpened edges must be perfectly straight, and the two knives must be exactly parallel to each other and level. To accomplish this type of leveling requires a precision machinists level (see Chapter 3).

For light propellers, the knives may be a couple of heavy steel rules, as their edge is probably straight and of the correct thickness (relatively sharp). Some method of adjustment must be given to the knives to allow them to be set parallel and level. If the propeller attaches to a propeller hub that bolts on to a tapered shaft or crankshaft, static balancing is usually accomplished with the prop hub on the prop.

With the propeller's arbor resting on the knives, there are no friction forces and the propeller is free to rotate in any direction. The heaviest blade will try and rotate to the bottom. By adjusting the weight of the individual blades, it should be possible to achieve a perfect balance such that the propeller has no tendency to rotate in any direction.

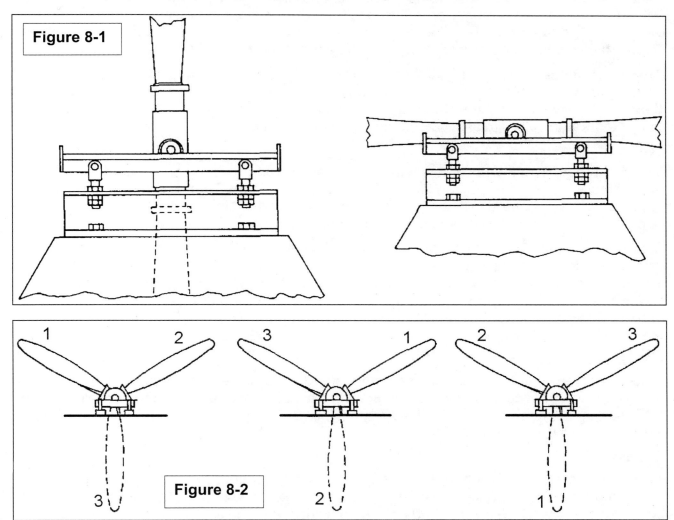

If the propeller is properly balanced statically, it will remain at any position in which it is placed. Check two bladed propeller assemblies for balance, first with the blades in a vertical position and then with the blades in a horizontal position (Figure 8-1). Repeat the vertical position check with the blade positions reversed; that is, with the blade which was checked in the downward position placed in the upward position.

Check a three bladed propeller assembly with each blade placed in a downward vertical position, as shown in Figure 8-2.

An acceptable balance check requires that the propeller assembly have no tendency to rotate in any of the positions previously described. If the propeller balances perfectly in all described positions, it should also balance perfectly in all intermediate positions. When necessary, check for balance in intermediate positions to verify the check in the originally described positions.

Adjusting Static Balance

Weight that is to be added or removed in static balancing should be performed in accordance with the manufacturers' instructions. It is advantageous for the weight to be as far from the center of the propeller as possible to allow for the minimum change in weight. But, changes in the weight distribution of a propeller can alter it's response to vibration negatively, especially when the weight is close to the tips. This is more of a problem on a metal propeller. Also because there is less material at the end of the propeller, it is more difficult to alter the weight without changing the strength or stiffness. The type of propeller determines the method of weight change. Manufacturers' instructions will often dictate a minimum and maximum weight for any balance material, as well as for the whole propeller assembly.

There are three basic materials used in propeller construction; wood, aluminum, and composite (carbon, fiberglass, etc.). Many wood blades have a fiberglass layer to protect them, and some composite blades are of a sandwich construction that use a wood core laminated with the fiber (glass, carbon, etc.) to provide the necessary strength.

Wood

Wooden propellers may become statically unbalanced simply from being stored vertically, as the moisture they contain will tend to travel to the lowest blade. To help minimize this, park the airplane with the propeller blades horizontal. Before condemning a wooden prop as statically unbalanced, leave it for a week in a horizontal position.

Most static balancing can be accomplished by the addition or removal of some of the protective coating on the wood (varnish, epoxy/glass, metal leading-edge protection, etc.). It is easiest to apply some more varnish to the light blade. Adding glass fiber/epoxy sheathing can make a large difference in weight and is easily adjustable by sanding some off, but must be applied to bare wood to obtain a satisfactory bond (if it comes off in flight it is going to make the prop vibrate badly).

Aluminum

Static balance of aluminum blades may be altered by the addition or removal of paint. Some manufacturers provide instructions for removing metal from the tip of the propeller for the purposes of damage repair and/or balancing. The shape of the tip is important and all of the tips must be all the same.

Composite

There are a number of different types of composite structures used to make propeller blades. The manufacturers' guidance is the only safe way to alter the primary structure of a composite blade, if its' allowed. Like wood and aluminum blades, the addition or removal of finish (paint, varnish, epoxy) will accomplish some static balancing unless the propeller has been damaged.

Blade Tracking

Blade tracking is the process of making the propeller tip-path-plane perpendicular to the crankshaft. It makes the angle of attack of each blade constant throughout its arc, when the relative wind is parallel to the thrust line. It also makes the propeller axis of inertia coincide with the axis of rotation (see Chapter 7).

Blade tracking is easily measured but difficult to change. Blade tracking is measured by rotating the propeller by hand and measuring its distance from a point as the blade passes that point (Figure 8-3). All blades should be the same distance, ±1/16".

NOTE

Some manufacturers of wood or composite bladed propellers allow 1/8" tolerance in tracking. In addition there may be considerable allowable movement of the blade in the hub on the constant speed propellers. When tracking the propeller, hold the blades in the same position as they pass the reference point.

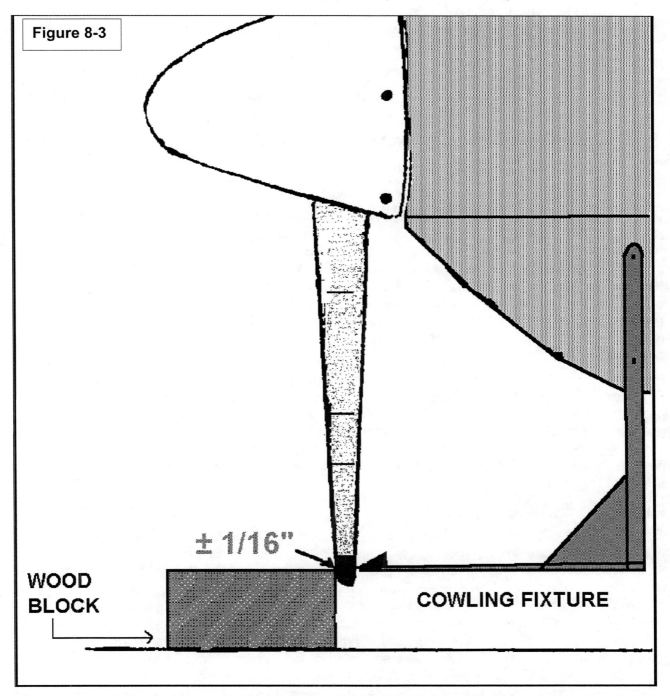

Figure 8-3

Accurate propeller tracking requires chocking the aircraft in a stationary position, removing the spark plugs so the engine may be turned easily (without disturbing the position of the airplane), and ensuring that the engine crankshaft is tight against a thrust bearing (push it or pull it but be consistent). In practice it is very difficult to rotate or push/pull the propeller without disturbing the aircraft unless the aircraft is off the inflated tires. Place a solid object (e.g. a heavy wooden block that is at least a couple of inches higher off the ground than the distance between the propeller tip and the ground) next to the propeller tip so that it just touches, or attach a pointer/indicator to the cowling itself (Figure 8-3). Scribe a line on the block next to the blade tip position. Pull all the blades past the scribed datum to determine their individual variation.

Improper tracking is generally the result of one or more things;

- There is foreign material (corrosion products, burr, etc.) between the propeller, spinner backplate, and crankshaft flange. These surfaces should be cleaned before assembling.
- The propeller bolt torque is incorrect.
- The spinner backplate is warped or otherwise defective. Check for a uniform thickness of the metal.
- Inconsistency between individual blade angles will affect the reading of the blade track.
- The propeller is bent or has some other defect.
- The crankshaft or crankshaft flange is bent.

Wooden propellers are subject to becoming loose on the crankshaft and out of track if not retorqued at regular intervals. They loosen due to the wood fibers changing under the pressure of mounting and moisture changes in the wood. Sensenich Wood Propeller Company publishes a detailed guide on the installation and maintenance of wood propellers, it can be downloaded from their website. The Ed Sterba Propeller Company also specializes in wood propellers and publishes detailed information on their website regarding the installation and maintenance of wood propellers. On wooden propellers (those with a wooden hub), the track may be affected by the order of torquing the propeller bolts, and this may be used to advantage to correct a slightly out of track condition.

Propeller tracking may also be checked with some types of dynamic balancing equipment, however it is typically not necessary unless one is trying to troubleshoot a very specific problem. It involves putting reflective tape on the propeller blades and watching them as they run using a stroboscope. This is commonly done on helicopters. The blades on airplane propellers shouldn't be varying that much from the static position, unless the propeller is very flexible, the individual blade angles are different, and the engine is developing substantial power at the time.

Dynamic Balancing

Notwithstanding the discussion on dynamic balance in Chapter 7, dynamic propeller balancing equipment can't resolve between the two basic kinds of imbalance, unequal mass and unequal mass distribution. In addition dynamic propeller balancing involves the addition or removal of weight along one plane (the spinner backplate), so it cannot rectify an unequal distribution of mass. It is called dynamic balancing simply because it is done while the powerplant is operating. Even though it is referred to as propeller balancing, it is really attempting to balance the whole combination of the propeller, spinner, and engine, since the three are rigidly interconnected. Because of this it is possible to obtain a remarkable difference in noticeable vibration. In addition, very small differences in weight between the blades that wouldn't appear in static balancing result in a vibration that increases with RPM and can be measured with dynamic balancing equipment.

Dynamic balancing will reduce the overall powerplant vibration and increase performance, at the power setting that balancing was performed at. Dynamic balancing requires specialized equipment that is not available at all FBO's performing maintenance (Figure 8-4).

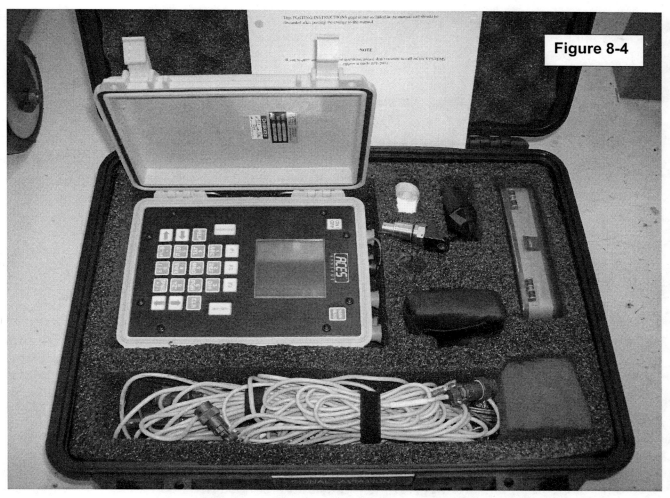

Figure 8-4

Sometimes it can be rented through the mail. The equipment is not difficult to use and the manufacturers of the equipment write detailed instructions for the mechanic (the website for ACES Systems provide some excellent tutorials and instructions). However, the equipment is sophisticated and can be intimidating if one hasn't been shown before. There are also subtleties to efficient balancing that an experienced technician can help with.

Prior to performing dynamic balancing, the engine should be running smoothly without objectionable vibration. Dynamic balancing cannot be performed on a powerplant that is vibrating excessively from;

- a poor suspension system
- a statically unbalanced propeller
- an untracked propeller
- a propeller with unequal blade angles
- a rough running engine.

If the analyzer could measure the excessive amplitudes of the vibration from the above cases, the amount of weight that would need to be added might cause structural problems. Dynamic balancing is the last step in propeller/powerplant rigging and everything must be properly adjusted and in good working order prior to getting there.

The typical setup for dynamic balancing uses two sensors (accelerometer and photo-sensor) and a computer. The accelerometer measures the vibration and is mounted as close to the propeller as possible, on the engine. The photo-sensor looks at one propeller blade for timing.

NOTE

Some systems allow using a second accelerometer, installed at the rear of the engine case, to increase accuracy.

The engine is run at the desired RPM for balancing and the computer will calculate exactly how much weight is needed, and exactly where it is to be put. Typically the weights are just regular AN washers attached to the spinner backplate, and is often on the order of a few grams. The computer provides an angular measurement of weight location, measured from the propeller blade with the reflective tape, around the periphery of the spinner, for example, 3.2 grams at 276°.

NOTE

Some early dynamic balancing systems require the user to manually plot the readings on a graph provided by the manufacturer, to calculate the amount and position of weights.

Airplane propeller dynamic balancing measures amplitude in velocity, Inches Per Second. Generally speaking, the vibration coming from the front of a reciprocating engine should not have amplitude greater than 0.20 IPS. A vibration of greater than 0.20 IPS is indicative of other problems and can't be rectified through dynamic balancing. The balancing computer will output the actual velocity throughout the balancing process. If everything is working well, it is possible to get the vibration down to .05 IPS through several iterations of dynamic balancing. This will produce a noticeable reduction in vibration and an increase in performance. Dynamic balancing should be the last step performed in airplane rigging.

Engine

Like other aspects of airplane rigging, the manufacturers' documentation should be consulted where available. Because of the wide variety of engines, only a few general things are discussed here.

Idle Speed

The idle speed of the engine must be low enough for the airplane to land. On a small low-wing airplane with a relatively high aspect ratio wing, an idle speed of more than 800 RPM may allow the airplane to maintain flight in ground effect nearly indefinitely.

Mixture

The mixture of most air-cooled airplane engines is set slightly rich at idle, such that one will see an increase in engine speed of about 25 RPM when the mixture is retarded slowly to the idle cut-off position.

Controls

Like the primary flight controls, the engine controls should be adjusted so that the controls (mixture/throttle/wastegate arm) touch their stops (Figure 8-5) slightly before the engine controls in the cockpit hit their stops (or run out of travel). This usually about 1/16" – 1/8" of travel left at the cockpit control (Figure 8-6).

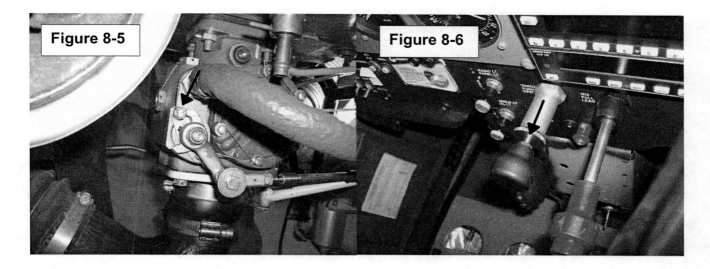

Figure 8-5   Figure 8-6

It is common to provide a light spring on the throttle and mixture arms so that if the linkage becomes disconnected, the throttle will advance to full power or the mixture will go to full rich, rather than cause an engine failure.

Turbo/Supercharger

In the absence of automatically adjusted boost pressure, it is desired that slightly too much be available at full throttle. The ideal situation is that it is possible to obtain takeoff RPM and boost with *nearly* full throttle at sea level, and that the last bit of throttle provides a slight excess of boost pressure. Most turbo/supercharged aircraft require some discipline in throttle pushing to make sure that no overboost occurs, even turbonormalized engines with automatically regulated boost pressure. If the system is provided with an adjustable overboost-relief-valve or other device for regulating maximum boost, some adjustment may be necessary after it has been placed in service for a time.

Propeller and Propeller Governor Settings

Reciprocating engine powerplants must make some minimum static RPM (achievable RPM on the ground at full power) to have acceptable takeoff and climb performance. The actual amount it is supposed to be is a fundamental part of the aircraft design, and is usually stated in the Pilots' Operating Handbook on certificated aircraft (it is a method of determining if the engine is developing normal power). If the combination of engine and prop are known to work and the static RPM is too little;

- the engine is not running properly, or
- the tachometer is incorrect, or
- the blade angle/propeller pitch is too great, on an adjustable prop, or
- the low pitch stops on a constant speed prop are set incorrectly (not usually the case), or
- speed setting on the governor of a constant speed prop is set incorrectly, or
- the boost pressure is too low.

RPM checks should be done in relatively calm air (less than 5 MPH wind).

Adjusting Ground Adjustable Propellers for Performance

A propeller whose pitch does not vary in flight is a compromise between an acceptable cruise speed, and acceptable takeoff/climb performance. At a given power setting, the RPM of an engine with a fixed pitch propeller varies with airspeed. At low airspeeds, the propeller moves a greater amount of air and hence has a load which reduces the attainable RPM. At higher forward airspeeds, the inflow to the propeller is partially a function of the forward velocity of the aircraft, which reduces the propeller load and allows the engine to attain a higher RPM.

A propeller with a lesser blade angle will allow the engine to develop a higher RPM for any given power setting. The higher RPM at low speed equates to increased static thrust, advantageous for takeoff and climb. Because it will reach the maximum allowable RPM sooner, the maximum thrust is reached sooner as airspeed increases, and the airplane will have a lower top speed.

A propeller with a greater blade angle will bog the engine down to a lower RPM for any given power setting. The lower RPM at low speed equates to decreased static thrust, reducing takeoff and climb performance. Because it will reach the maximum allowable RPM later as airspeed increases, the aircraft accelerates longer and the airplane will have a higher top speed.

The engine/propeller RPM is not just limiting from a mechanical standpoint, the propeller tips must operate below the speed of sound to be efficient (wooden propellers may shatter at supersonic tip speeds). The speed of sound at sea level is around 1,100 Feet per Second. On a given propeller, some higher RPM will cause the tips to approach the speed of sound. The diameter of the propeller is also a factor since a propeller with a larger radius must travel further through the air on a given revolution (Figure 8-8). A larger propeller then must be turned at a lower RPM to keep it subsonic. The actual tip speed is generally kept below about Mach .8 (about 880 FPS) because as the speed approaches supersonic, some parts of the propeller may see supersonic flow due to the airfoil curvature, even when the tip speed is below 1,100 FPS. The noise gets really bad as this limit is approached. The forward velocity of the aircraft also affects the RPM where the tips become supersonic because the airspeed seen at the tips increases with aircraft forward speed.

Pitch is used to describe blade angle because it more easily represents the behavior of the propeller. The actual angle of the propeller blades vary along their length (Figure 8-7).

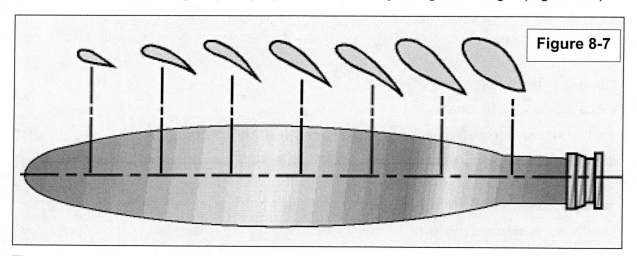

Figure 8-7

The outer portions of the propeller are traveling faster than the inner portions (Figure 8-8). To maintain a constant angle of attack along the propeller blade as it rotates, the blade is twisted.

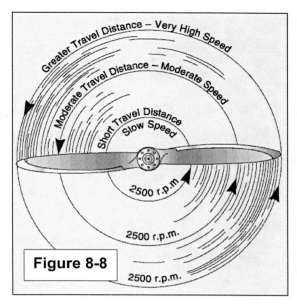

Figure 8-8

It has a greater angle inboard where it is traveling more slowly, and a smaller angle outboard where it is traveling much faster.

Propeller pitch is the linear distance the propeller would advance in one revolution if it were traveling in a solid medium, much like a screw going into wood. It is measured in inches. The pitch is constant along the span of the propeller blade, unlike the blade angle.

High pitch is the same as high blade angles and results in low RPM. Low pitch/low blade angle equals high RPM.

Pitch is not handy for setting the blade angles so manufacturers will specify an angular measurement at some point on the propeller from which to make measurements. The standard position is 0.75 radius of the propeller (Figure 8-9). Trigonometry is used to convert pitch-inches to blade angle-degrees;

$$\arctan A = P/C$$

Where;

A = angle of propeller blade at 0.75 Radius, degrees

P = Pitch, inches

C = Circumference of circle at 0.75 Radius, inches

- the circumference of a circle = $diameter \times \pi$
- the arctan or atan function is often symbolized by **tan^{-1}** on a scientific calculator (make sure the calculator is set to display degrees instead of radians).

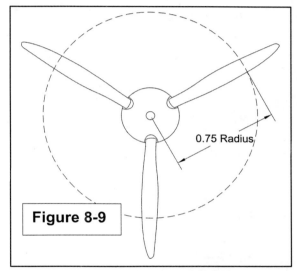

Figure 8-9

On ground-adjustable propellers, a change in *pitch* of one inch will result in approximately 100 RPM change in static RPM, at maximum power on a naturally aspirated engine.

Chapter 9
Aeroelasticity and Control Surface Mass Balancing

Flutter is a kind of forced vibration and the basic elements of vibration (stiffness and mass of the structure) are factors of flutter. A review of Chapter 7 may be helpful in understanding the terminology used in this chapter. Flutter is different from forced mechanical vibrations however, in that with flutter, the forcing functions (lift and drag) are dependent on the motion of the oscillating structure. As the structure oscillates, the lift and drag are constantly changing, interacting with the elastic structure and mass distribution of the aircraft to sustain the oscillations. Flutter then is a self-exciting vibration. Figure 9-1 illustrates the physical relationships.

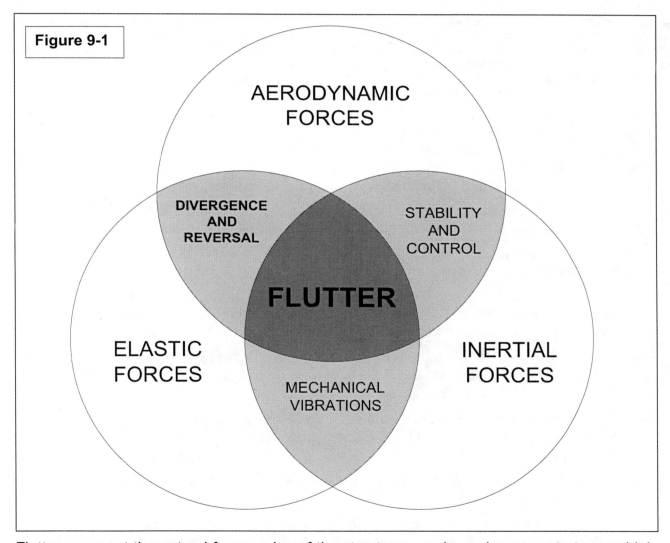

Flutter occurs at the natural frequencies of the structures, and may be present at some higher multiples of the natural frequencies, however, typically only the lowest frequencies are the ones that are important.

When flutter occurs, the amplitude of the oscillations may increase rapidly to destruction (a sort of resonance). The rate that this occurs is commonly just a few seconds or less from initial onset to destruction.

There are various flutter 'modes'. Flexure (bending), torsion, combined flexure-torsion, control surface torsion, fuselage torsion, fuselage bending, symmetric/asymmetric, the list goes on and each flying surface of the aircraft is subject to one or more of these flutter modes. Some modes may become coupled together to produce even more flutter problems. Most small aircraft will only suffer from a few of the possibilities, the type and number depending on the stiffness of the structure and the speeds involved.

Formal analysis of the various modes is comprehensive and faster aircraft require more in-depth analysis. Analysis is done by a combination of mathematical predictions, mapping of the natural frequencies of the structure(s), and flight-testing while monitoring vibration response.

NOTE

Mapping of the natural frequencies is done by exciting the structure(s) with vibrators and measuring the response of the structure(s) with accelerometers and/or strain gauges. This provides an indication of what vibrations are conducive to flutter. An aircraft is also flight tested with the same equipment installed to observe at what speed the vibrations start to become sustained by the interaction with the airloads. Because the difference between a well damped vibration and a divergent one (destructive flutter) is only a few knots in many cases, flight testing for flutter is still a hazardous test, even in well instrumented aircraft.

Flutter is dangerous and there are few quantifiable generalizations that provide protection from flutter in aircraft design. Nevertheless, References 1, 13, and 18 provide some simple guidance on design criteria for flutter prevention. The rigger is mainly concerned with the mass balancing of the control surfaces, and eliminating free play from the control systems. Some of the characteristics of flutter are given here.

- Because the aerodynamic loads increase with speed, it can be said that at some higher speed a particular aircraft will begin to suffer from flutter, or is more likely to flutter.

- An increase in altitude reduces the viscous damping effect of the structure and thus may cause flutter to occur more readily than at lower altitude.

- A stiffer structure will not suffer flutter as readily as a structure that is more elastic. Stiffness also relates to the control system (discussed in other chapters). Because of flutter, stiffness sometimes dictates the design of the structure, rather than strength.

- The distribution of mass on a particular structure may induce or prevent flutter. This includes the presence, or lack of, fuel and or fuel tanks near the wing tips, or unconventional mass distributions.

- Control surfaces that are mass balanced are less likely to suffer from flutter, or rather will flutter at a higher speed than unbalanced controls. Mass balancing is the subject of this chapter.

- T-tails, V-tails, boom type fuselages, and other unconventional configurations are often more conducive to flutter.

- Spring-type devices in the control system can both prevent or cause flutter depending on the aircraft. Some aircraft have springs to assist or hinder elevator movement, or use springs as a force trim system.

- It is sometimes given as a rule of thumb that aircraft capable of speeds in excess of 150 KTS need to have static mass balancing of the control surfaces, while aircraft capable of speeds in excess of 260 KTS need to have dynamic mass balancing of the control surfaces (the difference between dynamic and static balancing is discussed later in this chapter). These are approximations and aircraft have been destroyed by flutter at much lower speeds than those given here. On the other hand, there are homebuilt aircraft that go 200 KTS without any balancing, because their structure is very stiff.

- The rules given above are not comprehensive and flutter is affected by many other factors like control surface trailing edge condition, control surface shape, etc.. Much research is available on this subject.

The physical phenomenon is most easily illustrated by an example of flexure wing flutter (wing bending) with an unbalanced aileron, as in Figure 9-2. At position A, a disturbance causes upward bending of the wing. If some small free play exists in the aileron hinge or aileron control circuit, the aileron will be deflected downward relative to the wing as in position B (Figure 9-3). This is because the center of mass of the unbalanced aileron is behind the hinge point, therefore its reaction to a sudden movement is to stay where it was (because of inertia). The airloads now on the deflected aileron drive the wing up further. At C the stiffness of the structure finally stops the movement and the elastic restoring forces of the wing structure begin forcing the wing back downward. At D, the aileron has again deflected in such a way to make the airloads accelerate the downward travel of the wing. This time the airloads are added to the downward acceleration due to the elastic restoring forces, and inertia helps to drive the wing even further downward (point E) than it went upward in the first half of the cycle. After E, the process is repeating. This example illustrates a divergent oscillation in which structural failure will eventually occur. Like mechanical vibrations, the amplitude increases rapidly as the frequency of the oscillations approach the natural frequency of the structure. The difference in speed between a no-flutter condition to a divergent oscillation is often only a few knots.

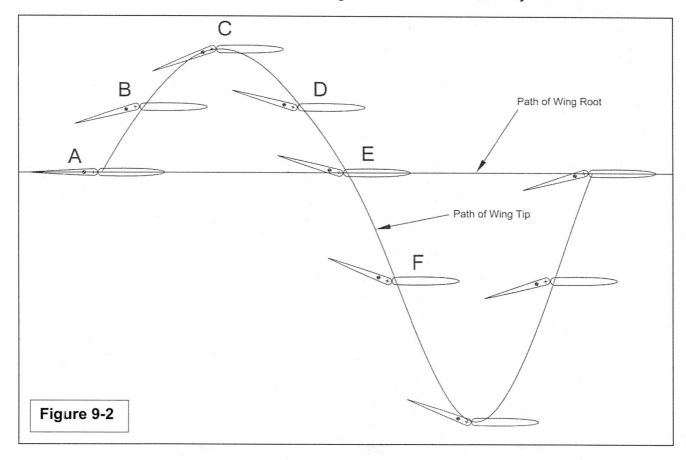

Figure 9-2

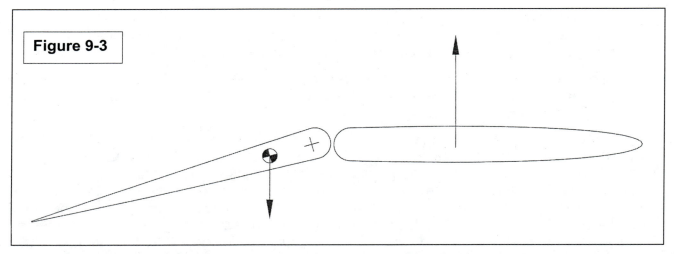

Figure 9-3

If the aileron were balanced such that its' center of gravity was at its hinge line, it would not lag behind the acceleration and bending of the wing, and it would not contribute to the divergent oscillation. The oscillation would die out quickly as the wing will have only been demonstrating free vibration.

This example is a very simple case and flutter is usually more complicated. A more common case of aileron flutter is as above, but the wing twists in addition to bending, increasing the adverse airloads on the wing/aileron, and hence the amplitude of the oscillations.

Mass Balancing

There are two types of mass balancing procedures performed on control surfaces, static and dynamic.

Dynamic mass balancing is not to be confused with *aero*dynamic balancing of control surfaces, used to lighten control forces at high speeds. The aerodynamic balance in Figure 9-4 is often the location of the mass balance weight.

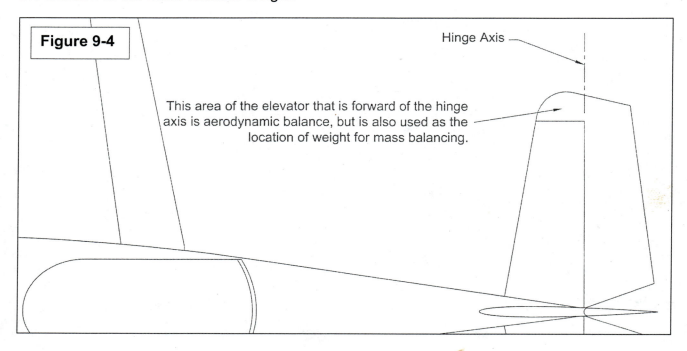

Figure 9-4

This area of the elevator that is forward of the hinge axis is aerodynamic balance, but is also used as the location of weight for mass balancing.

Hinge Axis

Static Mass Balancing

Static balancing of control surfaces is a distribution of weight *about the hinge axis (on either side of it)*, such that the control surface will balance on its hinges, in a level attitude as in Figure 9-5 (the center of gravity is at the hinge line and the chord line of the airfoil is approximately level). This keeps the surface from lagging behind the motions of the supporting structure that are acting perpendicular (or approximately so) to the hinge line of the control surface, illustrated in the example in Figure 9-2.

Dynamic Mass Balancing

Dynamic balancing is a distribution of balancing weight *both about and along the hinge axis* of the control surface. The theory of the dynamic balance here is similar to that discussed in Chapter 7, in that the axis of inertia does not coincide with the axis of rotation. The result of dynamic imbalance of a control surface is that a rotation or movement about some axis other than the hinge line will induce a rotation of the control surface about its hinge line. The flutter example demonstrated previously considered only the plunging action of the wing, i.e. a movement normal to the hinge line of the control surface. The solution to the example problem was to eliminate the inertial lag of the control surface by employing balancing weights to equalize the weight on either side of the hinge line. In the dynamic case, a bending or torsion about axes other than the hinge line can induce a reaction of the control surface that static mass balancing cannot prevent. For example, when the wing twists during oscillation and the elastic axis (twisting axis) of the wing doesn't coincide with the hinge line of the control surface (as is normally the case), the aileron will respond with movement even if it is statically balanced about the hinge line.

Like with mechanical vibrations, dynamic balancing also accomplishes the necessary static balancing. Dynamic balancing can be advantageous in that a complete balancing solution may require less weight than needed for a static balance (because perhaps the wing is much stiffer in bending than torsion, and isn't likely to suffer from the flexure flutter example given previously).

Dynamic balancing is accomplished by distributing the balancing weight appropriately along the hinge axis of the control surface, in addition to on either side of the hinge axis. Unlike the dynamic balancing of rotating machinery, the control surfaces are not rotated at high speed to determine their product of inertia. It is mainly a calculated value, based on the mass distribution of the control surface and the elastic axes of the supporting structure.

Determination of dynamic balancing of control surfaces is out of the scope of this book, however, when a particular design of aircraft is specific about the required position(s) and amounts of the balance weights, it may be for dynamic balancing rather than structural reasons. The reader is referred to References 13 and 18 for information on determining dynamic mass balancing requirements. The techniques given here for measuring the balance required are appropriate for either type of mass balancing assuming that the quantity and location of required weight is already known.

Balancing Terminology and Mathematics

Mass balancing of the control surfaces may be specified by the designer as positive, negative, or neutral.

A control surface which balances on its hinge axis in the same attitude as it would be in high-speed flight (chord line approximately level) is considered *neutral* or 100% statically balanced (Figure 9-5).

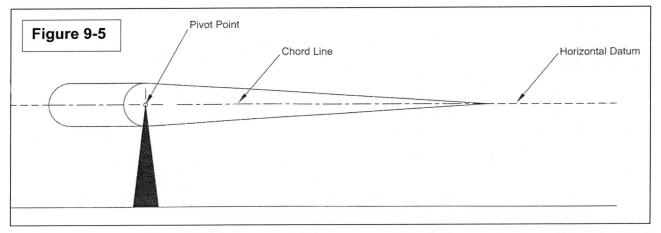

A balancing requirement specified as *overbalanced* indicates the control surface should be leading edge heavy/trailing edge light (Figure 9-6). This is referred to as *negative* balancing (the math is discussed in the next paragraph), or, as a percentage that is greater than one hundred (110% balanced for example).

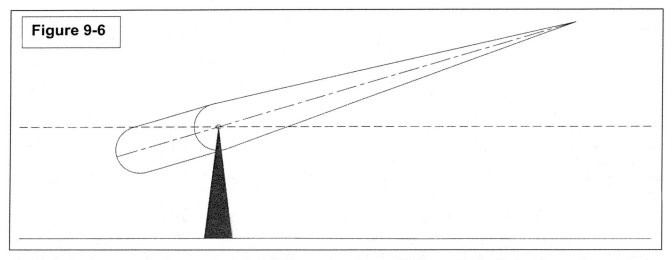

Underbalanced is leading edge light/trailing edge heavy (Figure 9-7). It is also referred to as *positive* balance, or, it may be specified as a percentage that is less than one hundred (80% balancing for example).

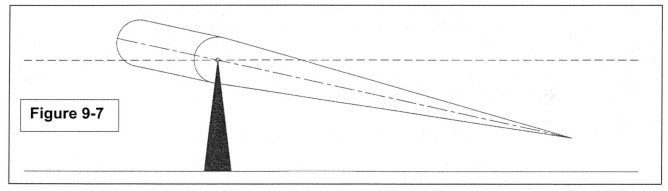

The math of balancing the control surface is essentially the same as balancing the whole aircraft about its center of gravity. The weight of an object multiplied by its distance from a given point is a measure of the amount of force that attempts to rotate the object around that point. There are three variables in the equation;

- *weight*, considered in ounces for this manual,
- *arm*, the distance of the weight from the hinge line, and measured in inches in this manual,
- *moment*, which is the product of the weight and arm (*weight* X *arm* = *moment*), and is a measurement of torque. Because inches and ounces are used in this book, the unit for torque is inch/ounces (abbreviated in/oz).

An unbalanced control surface has the majority of its' weight aft of the hinge line. This is a moment about the hinge axis that attempts to rotate the control surface trailing edge down. That moment can be calculated by weighing the trailing edge of the control surface while it is suspended on its hinge line and free to rotate about it; and multiplying the weight seen on the scale by the distance from the hinge line to the trailing edge (point where it was weighed). An example is given.

It is typical in aircraft maintenance manuals that the negative direction of the arm is forward of the hinge line and positive numbers are aft of the hinge line. In the example in Figure 9-8, the weight is 3 ounces at +12 inches, creating a moment of +36 oz/ins. To balance the control surface 100%, a torque of -36 oz/ins must be applied. The negative number arises from the arm measurement, any point forward of the hinge line is a negative arm. To evenly balance the +36 in/oz, there could be 12 ounces at -3 inches, 18 ounces at -2 inches, etc., any product that makes -36 oz/ins.

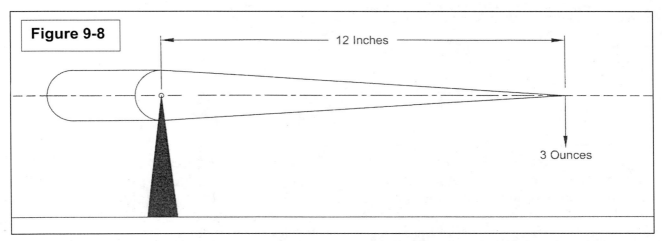

It is common for balancing requirements to be specified as something other than perfectly balanced. Quantitatively this may be expressed as a percentage (110% balanced for example), or a moment (-3 oz/ins for example).

If a specification calls for some percentage of balance, the positive moment is multiplied by the specified percentage and that amount of weight is placed ahead of the hinge line. For example, a specification for 110% balancing in the previous example requires 39.6 oz/ins forward of the hinge line (36 oz/ins, multiplied by 110%), making the control surface overbalanced. A requirement for 80% balancing would be 36 in/ozs multiplied by 80% is 28.8 oz/ins, placed ahead of the hinge line to make the control surface underbalanced.

If a specification calls for the control surface to be balanced by some specific moment (+3 oz/ins for example), remember the positive numbers are aft of the hinge line and the negative numbers are forward of the hinge line. Using a desired balance of +3 oz/ins for the example, the positive sign indicates trailing edge heavy. Using the previous example in Figure 9-8, if the trailing edge should be heavy by +3 oz/ins, then the weight to put ahead of the hinge line is 33 oz/ins. If the specification called for –3 oz/ins, the weight to put ahead of the hinge line is 39 oz/ins.

For most small aircraft, it usually safer to balance or overbalance slightly (110%) where balancing is required and no specification or tolerances are given. Of course some minimum strength and stiffness must exist at the attach point of the balance weight.

Measuring the Balance and Attaching Weights

Before balancing, the control surface should be in its' final condition; painted, and any peripherals attached to it like trim tab(s), control horns, etc. Include trim tab actuating parts that are wholly supported by the control surface, they increase the weight of the control surface and must be accounted for. Do not include pushrods or other control system parts that are in some way connected to the primary structure. Balancing must be performed in a draft free room.

The reference point for balancing is for the chord line of the airfoil of the control surface to be level, unless otherwise specified (Figure 9-9). Even though the instructions may call for underbalance or overbalance, the calculations are made relative to the amount of weight that is required to balance the control surface such that the chord line is level.

NOTE

The precise attitude for balancing is actually the same attitude that the wing section will take when the aircraft is in high-speed flight (relative to the oncoming air). This is generally close to the chord line and the maintenance instructions for most aircraft call for the chord line to be level, rather than at some angle.

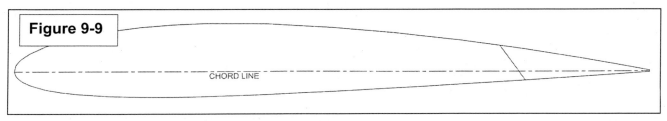

Figure 9-9

A setup like the one illustrated in Figure 9-10 will make balancing easier and accurate. The balance beam is easily fabricated from some aluminum channel and tubing. Alternatively, one can simply keep adding weights to their permanent location until the control surface balances properly. If the specification calls for under or overbalancing, the weight is removed and weighed to determine the actual balancing requirements. This method does require the ability to determine when the control surface is level by measuring up from the table to the chord line of the control surface, The balancing beam in Figure 9-10 is set parallel to the airfoil chord line on installation (Figure 9-13) and so provides a convenient reference from which to measure the orientation of the chord line.

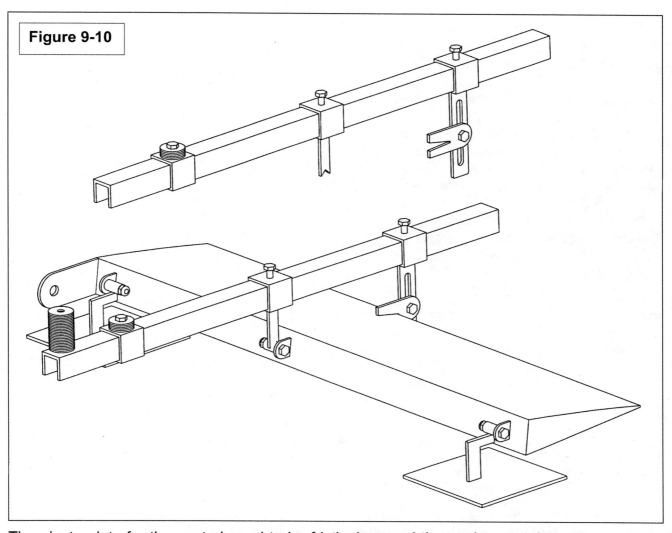

Figure 9-10

The pivot points for the control need to be frictionless and the easiest way is to use a couple 'knives', similar to propeller balancing (Figure 9-11). These are easily fabricated from scrap steel, the only precision required is that the two knife edges be approximately parallel and level along the knife edge and close to the same height. The hinge bolts are inserted into the control hinges and tightened, using round bushings if necessary for the hinges shown in the illustration.

The balancing beam should be set perpendicular to the hinge line for balancing unless otherwise specified, as in Figure 9-12 (as opposed to parallel to the wing chord line like when measuring control deflections).

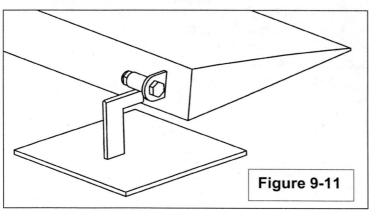

Figure 9-11

Once the beam is adjusted to fit that control surface, the beam itself needs to be balanced before taking measurements. Mark the locations of the beam setup and remove it. Balance the beam alone at the pivot point by shifting the adjustable weight back and forth.

Reattach the balancing beam in the same spot that it was. A weight of a known quantity is then placed on the beam and moved back and forth until the two ends of the beam are the same height above the surface on which the whole setup is placed (Figure 9-13). Alternatively, one may use a level on the beam as long as it is centered perfectly over the balance point (hinge line) and will not affect the balance Calculate the required moment to balance level and then apply whatever compensations are required (80%, -5 in/oz, etc.).

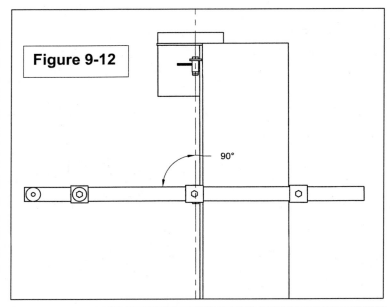

Figure 9-12

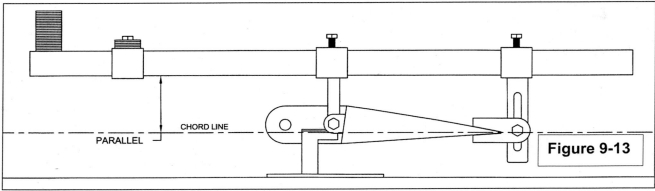

Figure 9-13

An alternative method of determining the amount of weight to use is to weigh the trailing edge of the control surface with a very sensitive scale while it is on the balancing knives in the proper orientation. The moment then is the measured weight, multiplied by the distance from the hinge line to the point of contact of the trailing edge to the scale.

There are different methods and locations of attaching balance weights. Lead is very efficient and is easily cast into a desired shape. It is generally easier to make a weight slightly heavier than necessary and then remove a little weight at a time to get the desired amount.

The strength of the counterweight assembly should be able to withstand the loads shown in Figure 9-14, for their attachment to the control surface. Those loads are easily obtainable at the if the control surface begins to vibrate or flutter and it has little to do with the flight load factors measured at the center of gravity of the aircraft. The purpose of balancing is to damp the vibration that is flutter, as well as prevent it. A control surface in certain flight envelopes may flutter and die out in one or two oscillations. However, the instantaneous acceleration of the control surface about its hinge line imposes a large load on the balance weight, requiring a great deal of strength in its attach point. If that weight were to separate during a vibration, destructive flutter may occur immediately

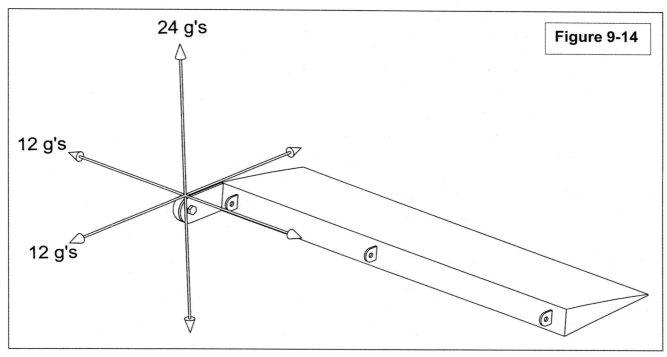

Figure 9-14

It is tempting to try and move weight further ahead to reduce the amount necessary, but this can cause structural problems. The attach point must be made very stiff or it may cause or contribute to flutter. Stiffness is a major concern in designing an aircraft structure to resist flutter. Designing a structure with the required stiffness to prevent flutter will often make it much stronger than is required. An example of one trick to avoid increasing the stiffness, and hence weight, is illustrated in the location of the mass balance weight in Figure 9-15. It is located at the midspan of a relatively long aileron so that the weight has less of a tendency to twist the aileron (aileron twisting may cause flutter, in addition to the weight imposing high torsional loads on the structure of the aileron).

Figure 9-15

Control surfaces must not be conducive to collecting water/ice that would upset the mass balance. This can happen in flight or on the ground. Either the control surface should be sealed or it should contain sufficient exit points to allow the water to pass through. It is normal practice to install drain holes in hollow structures regardless of the ability to resist water entry, to at least allow condensation to escape.

Chapter 10
Landing Gear Rigging and Vibration

Proper gear alignment makes the difference between being easy or hard to taxi, takeoff, and land. Poor alignment may result in difficult or dangerous handling, vibration, and increased tire wear. Most small airplanes are rigged with the same alignment geometry regardless of the type of landing gear, and that information is presented here.

Wheel Alignment

Alignment is performed with the aircraft in the ground rest attitude (all three wheels touching the ground), for both tailwheel and nosewheel aircraft. Rigging the landing gear on an aircraft with solid spring type landing gears, or the bungee type found on Piper high wings, requires that the aircraft be at its normal operating weight. This usually requires some ballast for the rigging procedure. Prior to rigging;

- Oleo struts should be appropriately serviced.
- Bungee type gears should have even bungee tension (no tendency for the aircraft to lean one way or the other).
- On all types of landing gears, all of the moving parts; linkages, pivot points, etc. should be lubricated at the hinge/pivot points.

Toe

Toe is the alignment of the main wheels with the longitudinal axis of the aircraft, as viewed from above (Figure 10-1).

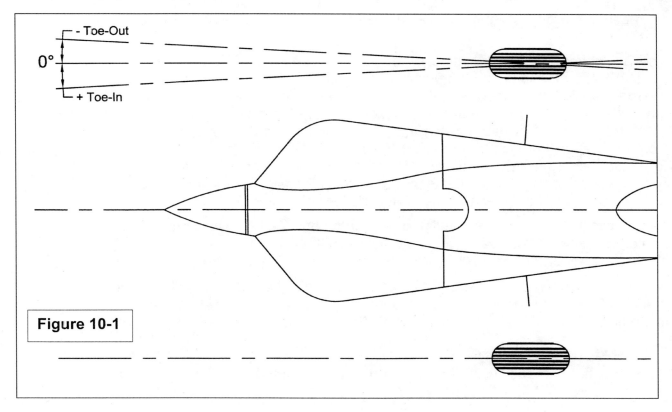

Figure 10-1

Toe-out is assigned a negative number and toe-in is assigned a positive number. Most instructions for aircraft will specify toe as simply toe-in or toe-out by a linear distance, for example, toe-in by a difference in distance of 1/8" between the front of the tires and the back of the tires (Figure 10-2).

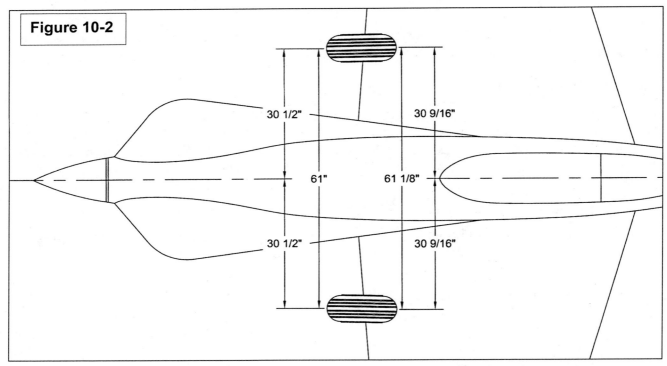

The main wheels of most aircraft are aligned with the longitudinal axis of the aircraft as viewed from above (zero toe). Some aircraft are designed to have a slight toe angle, but it is usually less than 1°.

Toe angle has the largest influence on ground handling. The effect of toe angle on any particular aircraft can have both positive and negative effects. It influences tailwheel aircraft to a larger degree than nosewheel aircraft. A small toe angle may achieve a measure of positive static stability for slight directional deviations, but cause problems with directional control for greater deviations. It depends on the gear geometry and mass distribution of the airplane. Some aircraft will benefit from a slight toe angle. Generally speaking, toe-out is destabilizing for straight line stability, while toe-in is stabilizing. Any amount of toe will increase tire wear because of scrubbing. The effects of toe (or other misalignment) won't be as noticeable when operating off grass or other strips with little friction.

Aircraft with solid spring type landing gears, or the bungee type similar to Piper high wings, can be very sensitive to non-zero toe angle because the rolling tires will attempt to force the gear closer together or farther apart (discussed again later in this chapter).

Setting the wheels to zero toe angle, at least initially, will generally provide satisfactory handling for either nosewheel or tailwheel aircraft. The toe angle is measured with the aircraft in the ground attitude (all three wheels touching the ground).

Camber

Camber is the perpendicular alignment of the wheels with the ground as viewed from the front or rear (Figure 10-3 and 10-4).

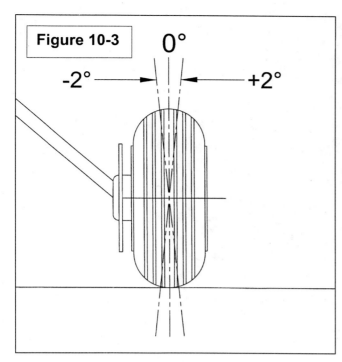

Figure 10-3

Aircraft are designed for approximately zero camber when the aircraft is on the ground at normal operating weight. With spring and bungee type landing gears, the camber changes with the weight (Figure 10-4).

By itself, camber plays little part in the directional characteristics of an aircraft, it mainly affects landing gear loads and tire wear. It may become coupled with other alignment problems in such a way as to magnify them. The tolerances for camber are generally larger than those for toe, as much as 5°-7° from the vertical on some solid gear aircraft. Oleo type main gears generally keep the camber within 1°.

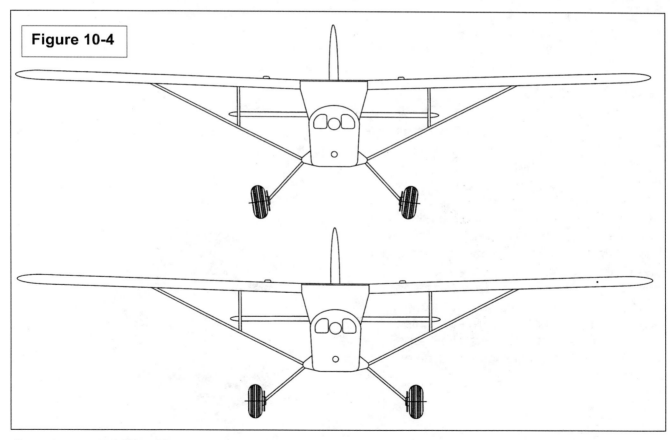

Figure 10-4

Caster and Trail

Caster is the angle between the ground and the rotational axis of the gear assembly (Figure 10-5).

The caster angle is usually a design or construction factor and is not normally adjustable, however, it is of sufficient importance when setting a gear or trying to troubleshoot gear problems that it is discussed here.

Caster angle combined with *trail* distance play a large part in the stability of an individual landing gear assembly. Trail distance is the distance between the castering axis of the gear assembly and the point of contact of the tire with the ground (Figure 10-6).

Caster angle and trail distance affect the stability and shimmy of steerable and castering nosewheels, and tailwheels.

Vibration in oleo-type main gears (Figure 10-5) may be traced to the caster/trail relationship when some free-play develops in the torque links, discussed later in this chapter.

Nosewheels

To provide good steering characteristics, the castering axis should be close to vertical (for a castering nosewheel), and some stability is provided by tilting the castering axis forward at the top as in Figure 10-6. To provide positive directional stability, the trail should be aft of the castering axis by some distance. These statements are very general and the reader is referred to Reference 20 for a thorough discussion of caster/trail relationships and their effect on stability and vibration. Some examples are given in Figure 10-7.

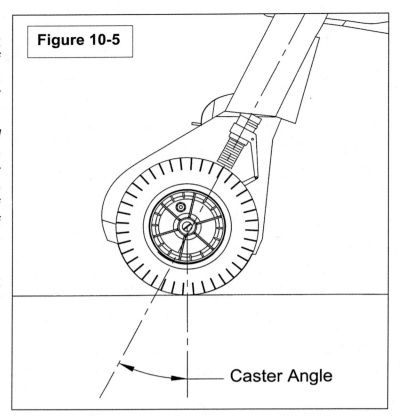

Figure 10-5

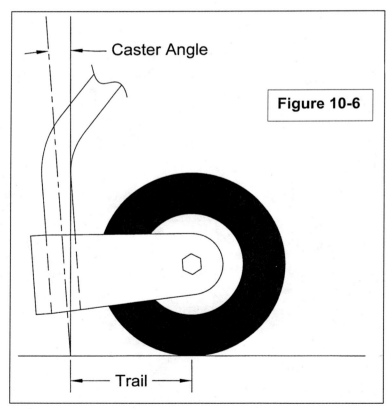

Figure 10-6

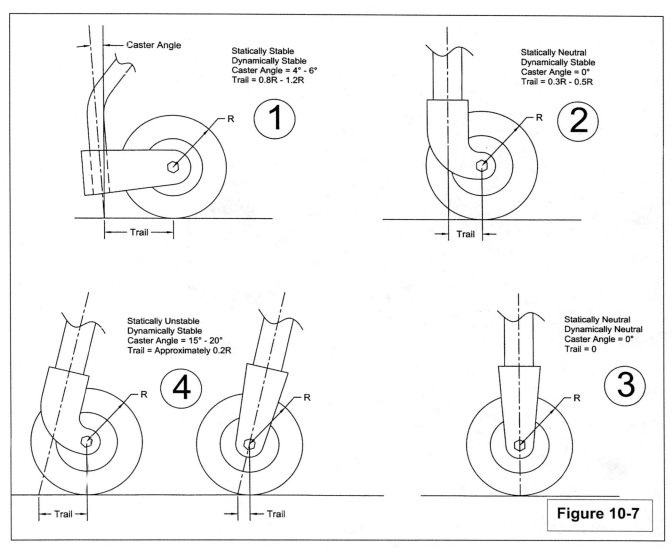

Figure 10-7

Because many small airplanes use castering or steerable nosewheels, the castering axis becomes a rigging issue if the aircraft is hard to steer, diverges directionally, or shimmies. Both stability and shimmy may be altered by changing the castering axis and trail distance. The trail distance is typically fixed by the design or part size, but some aircraft allow the caster angle of the nosewheel to be adjusted through the use of shims.

Tailwheels

On tailwheel aircraft, an adjustment in tailwheel height may be necessary to provide the correct attitude on the ground, especially when rearranging the incidences of the flying surfaces. This is often accomplished with shims between the gear spring and the fuselage (ensure that a hard landing is not going to force the tailwheel into the rudder).

The tailwheel castering axis must be vertical at the least operating weight (Figure 10-8),

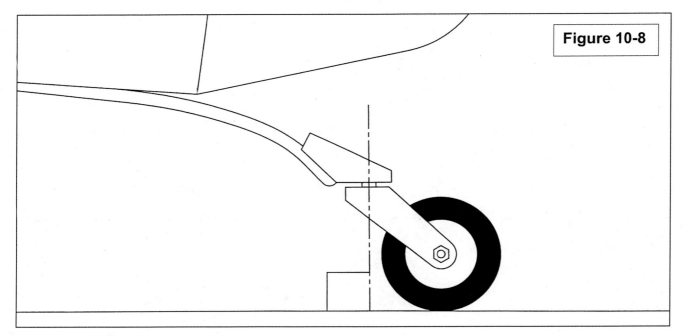

and tilt forward slightly as the weight is increased (Figure 10-9).

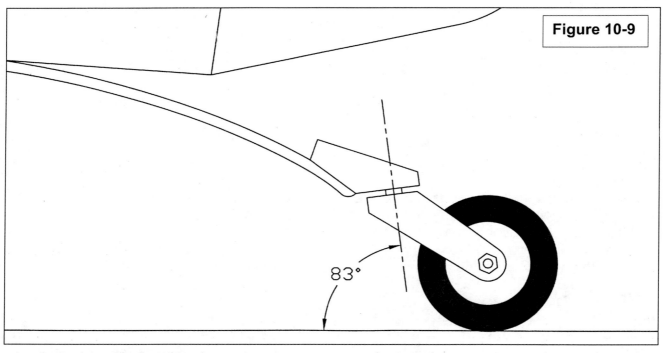

Tilt of the castering axis aft is going to create negative dynamic stability and difficult or impossible steering (Figure 10-10).

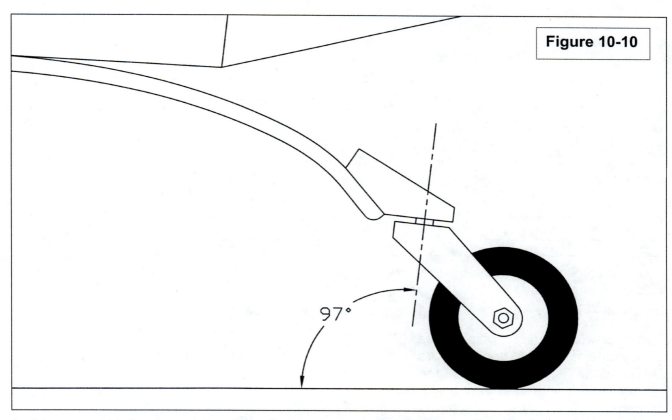

Figure 10-10

In this case the simple weight of the fuselage will force the wheel to one side or another when a turn is initiated. A tailwheel castering axis which is in the correct direction, but too large, will result in such great positive stability that it may be difficult or impossible to steer. The castering axis may be adjustable through the use of spacers or shims bolted between the fuselage and tailwheel spring, or tailwheel and tailwheel spring. Changing the overall height of the tail will change the takeoff/landing attitude and hence the takeoff/landing speed/distance and handling characteristics.

Rigging the Gear

The aircraft should be at normal operating weight, however, both light and heavy loads may be critical on some aircraft (in that the gear geometry changes with gear deflection). Some maintenance manuals provide the rigging specifications at empty weight, already having determined the change that will occur as weight is added.

Prior to gear rigging, it is necessary to draw some perpendicular lines on the floor over which the aircraft will be aligned. One line is parallel to the longitudinal axis of the aircraft centerline. The other line is perpendicular to the centerline and positioned adjacent to the main wheels (Figure 10-13). Plumb bobs are attached to the aircraft at the tail and firewall on the aircraft centerline to align the aircraft with the line on the floor (Figure 10-11).

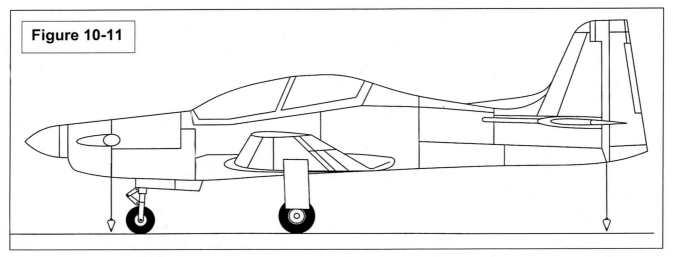

Figure 10-11

Use the following technique to draw the perpendicular line (a trammel bar will make a compass large enough).

To Erect at Point A, a Line Perpendicular to Line AB (Figure 10-12);

Using any convenient point C as center and radius CA, construct arc BAD. Draw line through BC and continuing to intersect arc at D. Construct line DA, which is perpendicular to AB.

To allow the gear to assume the position that it would have while moving, the aircraft is placed on greased metal plates, using ballast to load it down to its normal operating weight (Figure 10-13). This mainly important on solid spring or bungee type landing gears whose camber changes with deflection.

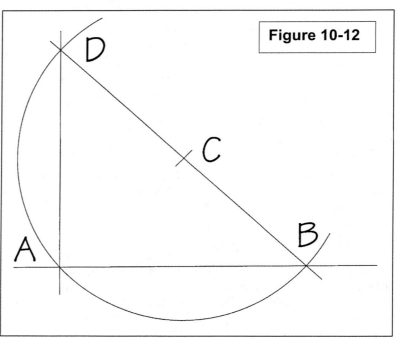

Figure 10-12

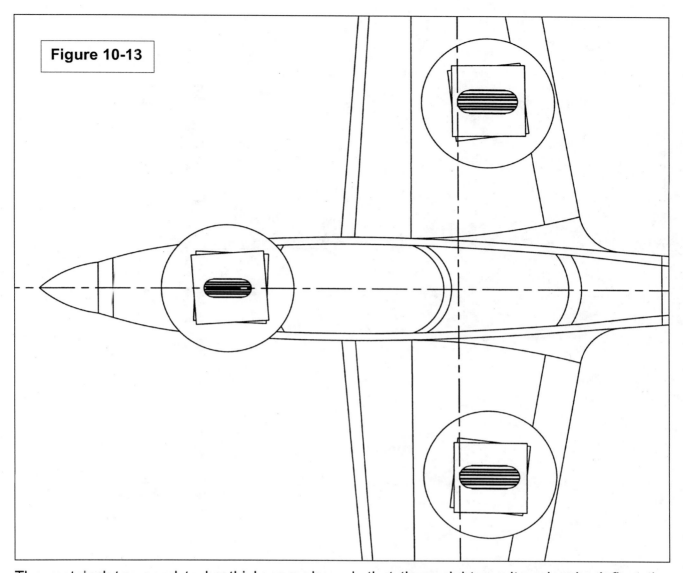

Figure 10-13

The metal plates need to be thick enough such that the weight won't seriously deflect the upper plate enough to create friction with the lower plate. This may be 1/8" thick for the smallest aircraft to ¼" thick for larger ones. A thick layer of grease is applied between the two plates and the aircraft is rolled or lowered onto the plates. Load the ballast and bounce the aircraft a bit to get the landing gear to assume the normal operating geometry. Realign it if necessary with the lines drawn on the floor.

With the airplane positioned over the centerline, the main wheels adjacent to the perpendicular line, and the aircraft leveled, toe, camber, and caster measurements may be accomplished with a square and long straightedge. Trigonometry may be used to convert a linear distance to an angle.

A long straightedge is used to span the two wheels, and is blocked up to be even with the centerline of the axles (Figure 10-14 and 10-15). A framing square is used (Figure 10-15) to align the straightedge with the line on the floor.

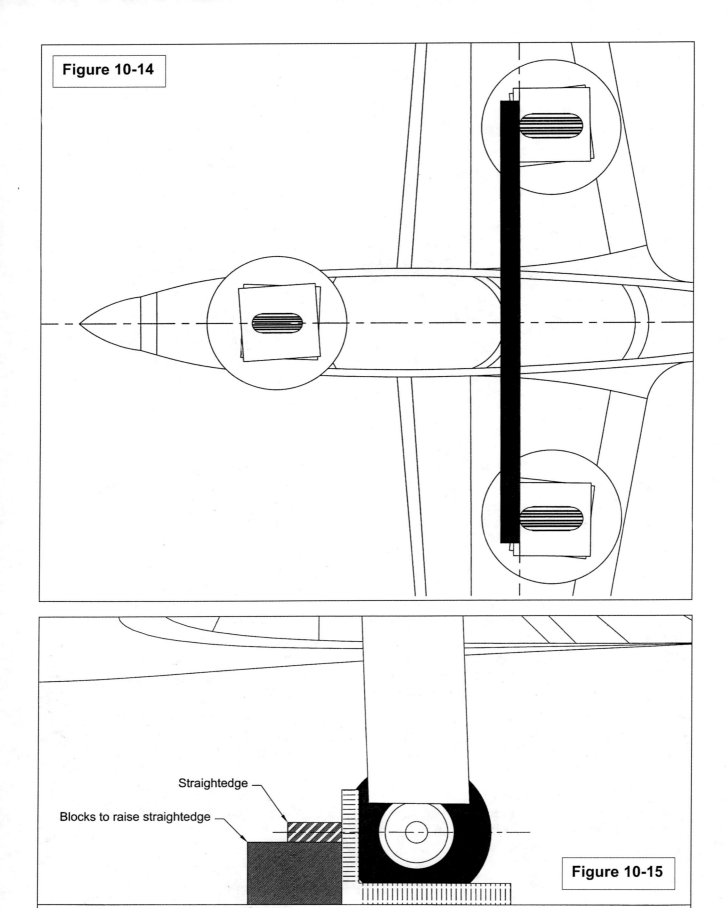

Figure 10-14

Straightedge
Blocks to raise straightedge

Figure 10-15

Any errors in caster of the main gear will show up here. Small errors (difference in forward/aft orientation of 1/8" between the two tires) aren't going to affect the aircraft very much, but any gross errors may affect directional stability at large deviation angles, and shimmy characteristics. If it's not possible to adjust the caster, make the straightedge touch the most forward wheel but keep it parallel to the line on the floor.

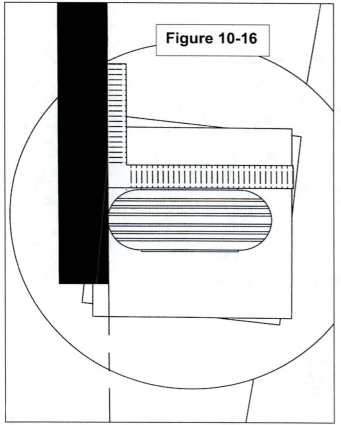

Figure 10-16

Toe is then measured by placing the framing square against the straightedge and the wheel flange as in Figure 10-16.

If the tire bulges out beyond the wheel, it is better to use some spacers against the forward and aft part of the wheel flange rather than put the square against the tire (Figure 10-17).

Trigonometry is used to convert a linear distance to an angular unit if needed.

Camber may be measured with a framing square on the floor but it is easier and more accurate to use a protractor like in Figure 10-18 (again, the aircraft must be level).

On solid spring type gears, toe and camber changes of the main gear are commonly made by using tapered shims between the axle assembly and the gear spring. On oleo type struts, toe is often altered with washers in the center of the torque links (Figure 10-19). Camber may or may not be easily adjustable, depending on the aircraft.

Caster on main gears is rarely adjustable for without disassembling major components. An incorrect caster angle on a main gear (one gear ahead of the other), is usually a construction or installation error, or it was bent during a hard landing.

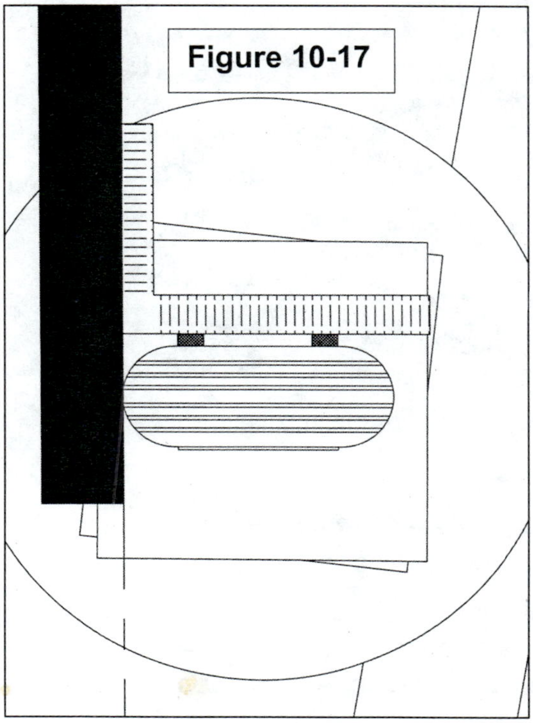

Figure 10-17

Figure 10-18

Steerable nosewheels and tailwheels should be adjusted while the aircraft is aligned with the reference lines. The pedals/rudder will have already been rigged prior to this (Chapter 5). The pedals are blocked in neutral position and the steering arm(s) and/or cables are adjusted to align the nosewheel/tailwheel with the longitudinal centerline on the floor. If the tailwheel or nosewheel is connected to the pedals through springs (Figure 10-20), spring tension should be equal among the two springs (it is necessary to raise the nose or tail momentarily between adjustments to allow the wheel to self-align under spring tension). Many tailwheel aircraft use springs between the cables and tailwheel steering arms that have no tension for small changes in direction of the tailwheel (they are loose when everything is lined up), but begin to apply tension at some amount of pedal deflection. Zero tension is sometimes used to prevent shimmy. On these aircraft the two springs should be slack approximately equally.

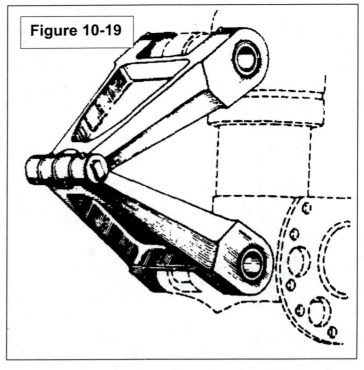

Figure 10-19

Figure 10-20

NOTE

Increasing tension to the springs in an attempt to increase stability or reduce shimmy may cause a loss of directional control when one spring breaks.

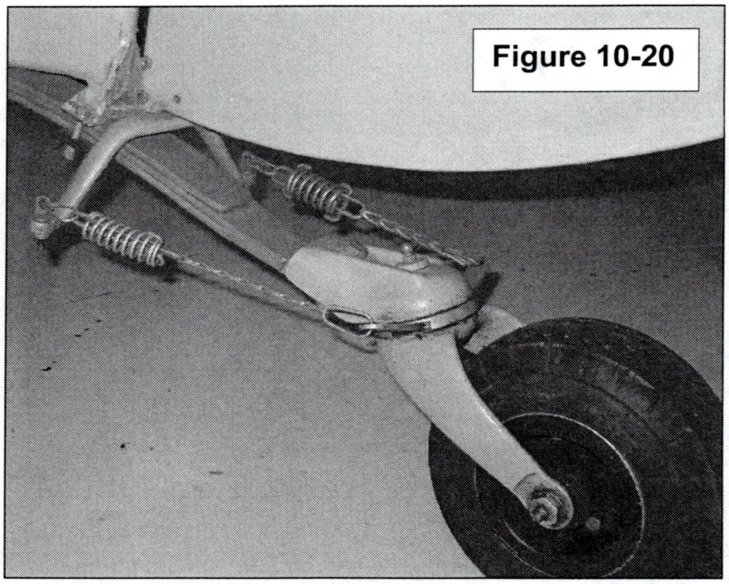

Landing Gear Problems

Landing gear problems are generally easy to identify and fix. Often times they aren't noticed until the aircraft is operated on a paved runway, when it had been operated predominately on grass. Much of the following discussion concerns vibration and shimmy as it is assumed the gear was rigged properly.

Check that the tire pressure is correct before taking things apart. Changes in tire pressure can affect the wheel/landing gear dynamics, including shimmy.

Wheels and Tires

Tire/tube manufacturers recommend that a new tire be allowed twelve hours at operating pressure before being put on the aircraft, so that the tire reaches full growth.

Make sure the wheel/tire was assembled properly. Tubeless tires should have the red balance mark adjacent to the valve. On tube-type tires the balance mark on the sidewall is the light spot, and should be adjacent to the balance mark on the tube, which is the tubes' heavy spot (usually the heavy spot on the tube is the valve). An improper installation can result in the tube being pinched by the tire, or the tube not inflating symmetrically (wrinkling). Ensure the tire bead is seated properly. The wheel halves may not have been assembled properly or in the correct orientation, or the wheel may be bent.

Small aircraft with low landing speeds don't usually get dynamic balancing of a wheel like automobiles, but static balancing may be appropriate. A balancing setup like the one used for static balancing of propellers in Chapter x will work for static wheel balancing as well.

NOTE

Static balancing of the wheel assembly is often not performed either on a small aircraft, unless there is a noticeable vibration.

A flat spot on a tire, usually caused by locking up a wheel during braking, may produce a noticeable vibration.

High frequency noise coming from a wheel may be a wheel bearing problem. An improperly preloaded or worn main wheel bearing may vibrate without producing noticeable noise (see the end of this chapter for instructions on preloading wheel bearings).

Brakes

Vibration may be caused by the brakes. A bent rotor will cause vibration with the brakes applied. Foreign material and corrosion products that exist between the steel disc and aluminum wheel during assembly may be the cause of vibrations. When servicing the wheels, the mating surfaces should be cleaned with an abrasive

Landing Gear

Notwithstanding the next discussions, the shimmy and vibration characteristics of landing gear are heavily influenced by their all-around stiffness.

Toe

A main gear which has its wheels toed in or out too much (even though equal in toe) can be troublesome, especially on spring gears. As the aircraft rolls, the main wheels attempt to roll in their own direction, to or from the aircraft centerline depending on the direction of toe. As the wheel travels, it slightly bends the spring gear until the force of the spring gear overcomes the traction of the tire, pulling it back suddenly. The frequency that this occurs at is generally low but of a large amplitude when it occurs. With small errors in toe, the tires simply scrub, wearing out quickly. If the aircraft has been operated on paved runways with an alignment problem, the tires may show uneven wear.

If the main gear isn't aligned properly with the direction of travel of the aircraft, it may be noticeable as a skipping or hopping at touchdown, similar to landing in a crosswind without making any corrections.

Camber and Caster

Errors in caster or camber of the main gears increases tire wear and imposes unwanted loads on the aircraft structure during landing. Gross errors in camber or caster may alter the other alignments when the aircraft assumes a different attitude.

As stated earlier, caster and trail affect the shimmy characteristics of pretty much all landing gears.

Wear

In some oleo strut landing gears and tailwheels (particularly the nose gears of some popular factory built aircraft), vibration or 'shimmy' is a common problem. It is usually the result of wear on more than one part making up the gear, although very worn or misadjusted wheel bearings will cause vibration alone. Hydraulic shimmy dampeners are usually not the culprit as they are not typically prone to failure (but the fluid might be low). Mechanical shimmy dampeners require occasional adjustment.

Most of the vibration problems in strut gears are caused by a little looseness developing in the torque links, shimmy dampener mounting points (where installed), and other parts subject to movement, through normal wear, in conjunction with a caster/trail relationship that is conducive to instability. Looseness may also be caused by improper assembly, having too much clearance in the components that are subject to rotation. Having slightly too much looseness, while maybe not causing vibration initially, accelerates wear very rapidly. A slight shimmy increases the rate of wear even more, causing looseness to occur in moving parts that would not normally wear from normal use. Wear is also accelerated by lack of lubrication.

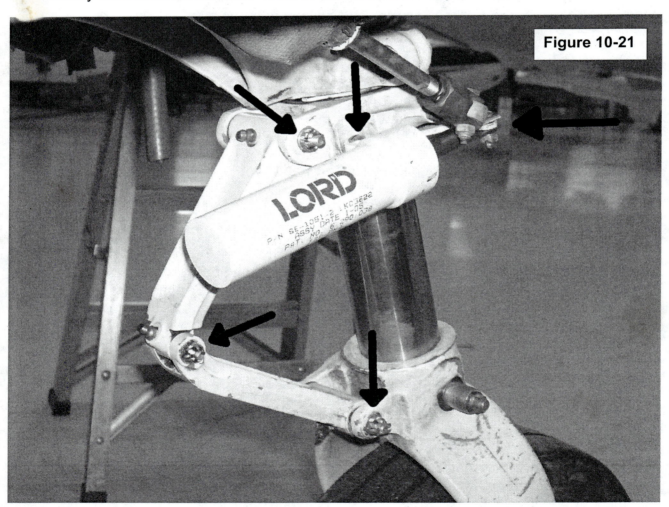

Figure 10-21

Tailwheel

If the tailwheel vibrates, first check for loose parts. The whole assembly gets a lot of abuse and develops loose parts after some time in service. It is necessary to support the tail of the aircraft on something besides the tailwheel or tailwheel spring to properly evaluate it for

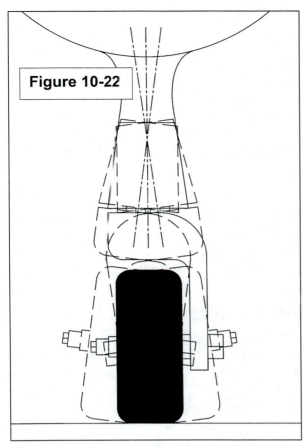

Figure 10-22

condition. A worn or unbalanced wheel may cause the shimmy, although most tailwheels are very small and any static or dynamic imbalance probably wouldn't be noticed.

Loosening or tightening the tension on the cables/springs, even slightly, may eliminate the shimmy. Increasing cable/spring tension may fix some problems, but it is common for tailwheel cable/spring connections to break, especially when operating off unimproved airstrips. If the springs are greatly pre-tensioned and one breaks, a loss in directional control may occur. They usually break during takeoff or landing. Certain tailwheel assemblies have different size springs between the right and the left. For clockwise engines (as viewed from the rear), the light spring goes on the left side.

Tailwheel shimmy can often be fixed by changing the caster/trail relationship. The tail wheel spring may be shimmed to make the castering axis approximately vertical when the aircraft is at its least operating weight. Also the weight imposed on the tail heavily influences the shimmy characteristics.

Some tailwheel assemblies use internal mechanical dampeners to prevent vibration. They are adjusted to provide a particular force necessary to rotate the tailwheel in castering mode. The increase in friction on the castering axis prevents shimmy. They require occasional adjustment.

The main tailwheel spring may be worn or inappropriate. A twisting motion of the spring (Figure 10-22) accompanies shimmy.

Larger tailwheels may need to be balanced, and sometimes a change in tire pressure of pneumatic tailwheels will eliminate shimmy.

Nosewheel

Nosewheel shimmy is a common problem on steerable nosewheels because they often have a caster/trail geometry that is conducive to instability. Most shimmy problems can be traced to worn components in the nose gear or steering linkage, although a dysfunctional shimmy dampener could be the problem (where installed).

Castering nosewheels, although usually having a geometry that is conducive to a certain amount of stability, are also subject to shimmy problems. Many castering nosewheel systems use an internal dampener (sometimes a simple mechanical device that increases friction) that provides friction on the castering axis to dampen shimmy problems. For small airplanes, the dampener provides a force of between 5 and 20 LBs that is required to rotate the nosewheel, with the force applied at the axle. Like tailwheels, a change in caster angle greatly affects the stability of the nosewheel.

Other Problems

A coupling between the spring rate of the landing gear and the response of a pneumatic tire sometimes produces a resonance which bounces the whole aircraft and makes control difficult. It is most obvious on landing for the low frequency modes, and for a stiffer setup may be set off by a rough surface at medium to higher speeds. On the tailwheel of such airplanes this manifests itself sometimes as an uncontrollable bouncing of the tail that occurs during a three point landing or when hitting a bump. Tundra tires have a much different response than regular tires and may lead to a lot of bouncing. Changes in tire pressure can change the coupling relationship.

Drum Brakes

Drum brakes may be found on some small experimental aircraft, antique aircraft, and some imported aircraft. Those that use a mechanical lever are prone to failure/jamming from mud, sod, or ice collecting on the lever arm at the wheel where the cable attaches to. In addition, an adjustment which provides some brake drag will cause overheating and possible locking of the brakes, often disastrous in a tailwheel airplane.

Preloading Wheel Bearings (Tapered Roller Bearings)

Assuming the wheel bearings have been cleaned, inspected, and greased; install the bearings and wheel on the axle and tighten the axle nut. Torque the axle nut enough to squeeze out the grease and seat the bearings (not more than 30 FT/LBS usually and not so much that it requires significant effort to rotate the wheel). Rotate the wheel backwards and forwards a few revolutions each direction. Loosen the axle nut, and then tighten it again only enough to get all the bearing surfaces seated. There will be a barely discernible friction on wheel rotation. Then tighten the axle nut just enough to get the cotter pin hole lined up. Wheel bearing looseness is evaluated by attempting to rock the wheel assembly back and forth (change in camber or toe), there should be no freeplay on these axes.

Chapter 11
Biplane Rigging

Biplane rigging varies somewhat between aircraft depending on the structural arrangement: whether the wings are one piece, two piece, or three piece; the configurations of the struts and wires; and the adjustments available. The service information, plans, and/or newsletters for a particular aircraft will provide more specific information than given here. Much of this section is taken from an old book on aircraft maintenance, written at a time when biplanes were commonly employed as working aircraft. Washout and incidence are used somewhat interchangeably here, and it is up to the rigger how to proceed in those respects depending on their particular airplane. This section provides sufficient information to determine the order of rigging for most biplanes. Many smaller biplanes and homebuilts do not have all the adjustments available of the larger biplanes, and this section will provide the necessary information to accurately locate the drilled holes for the various mounting points.

This section concentrates on the wings. Biplane geometry is discussed in Chapter 1. Empennage and landing gear rigging are given elsewhere in this manual, as well as general information on tie rods/flying wires, methods for making measurements, etc..

Level the Fuselage

Refer to Chapter 5 for lateral and longitudinal leveling prior to assembly.

Wing Assembly

The following paragraphs provide an outline of the assembly process. Rigging instructions are provided later in this Chapter.

The upper wing on many biplanes is made in three pieces. On these wings the small center section is mounted first. This will probably require more than one person unless a hoist is available. A larger biplane may require some platforms or scaffolding to stand on in order to rig the center section.

For small center sections it is easier to loosely mount the cabane struts on the center section while it is still on the ground. After the cabane struts have been attached to the center section, any fairings and inspection plates, which are to be installed over the struts, must be put in position or placed where they are readily accessible to slip over the struts before they are attached to the fuselage.

After the center section and cabane struts are secured to the fuselage, and the center section is in its' approximate position over the fuselage, the necessary wires or tie rods should be attached.

It is easier to precisely rig the center section before the wings are assembled, however, small errors in the rigging of the center section will result in large errors once the wings are assembled. It is still easier though to get the center section as close as possible to its exact location before attaching the wings, largely due to the number of degrees of freedom that are possible when rigging the upper wing.

When the center section is in place and rigged, the wings may be assembled. Whether the upper or lower wings are installed first depends on the mounting position of the upper ends of the landing wires. If the upper ends of the landing wires attach to fittings on the center section (the center section having already been installed), the lower wing is assembled first. If the upper ends of the landing wires attach to fittings on the upper wing(s), the upper wing is installed first. This is mainly to reduce the amount of jigging necessary to support the wings while they are being installed.

If the lower wing is to be installed first;

Hold the wing in position and insert the spar bolts (don't torque them yet). The landing wires are then installed to support the lower wings in their approximate position. Before actually attaching the wing, make sure that all wing connections such as aileron cables, lights, pitot tubes, etc., are connected or are in a position to be connected after the wing is installed.

NOTE

It is customary to install all tie rods so that the right hand thread is *down* or towards the *front*.

Attach the interplane struts to the upper wing (bolts loose) and make sure that all inspection plates, fairings, etc., which cannot be installed after the interplane struts are connected on both ends, are in place.

An adjustable rack to support the upper wing(s) as in Figure 11-1, may be necessary because of the time required to perform the rigging after all connections are made.

Figure 11-1

The cross bar should be about two feet longer than the chord of the wing and padded to protect the wing covering. The lower ends of the uprights should be anchored in some manner.

When lifting the top wing into position care should be taken not to allow the interplane struts to swing violently, possibly imposing a bending stress on the fittings or damaging the covering of the upper or lower wing. When the upper wing(s) are positioned, it is easiest to use undersize pins or bolts to temporarily fasten the wing. This will steady the wing but allow enough freedom of movement to permit alignment of all of the fittings. If there are any connections to be made from the top wing to the center section, such as aileron interconnect cables, electrical cables, etc., they should either be connected before the wing is attached permanently, or placed in a position where they can be connected after the wing has been attached. The actual spar bolts/pins and interplane strut bolts may then be installed. At this point they are not tightened in order that rigging adjustments be made later. After the permanent fastenings have been made, the wing support may be removed, and the flying wires installed. The top wing on the other side is assembled in the same manner. The wings are then ready to be rigged.

If the upper wing(s) is to be installed first;

The upper wing, with the interplane struts loosely attached, is installed on the center section and supported similarly to Figure 11-1. The upper ends of the landing wires may then be connected and tied roughly into position at the lower ends with a light string. This puts them in about the correct position and keeps them from interfering with the assembly of the lower wing. The lower wing is attached to the fuselage as previously described (with undersize pins or fasteners) and the interplane struts bolted in place. Landing wires may be attached next, the support removed, and the flying wires installed. The undersize bolts/pins are then replaced with the appropriate fasteners, but left untorqued until rigging is complete.

Wing Rigging

The general order of rigging given here applies to most biplanes and assumes the center section i;

1) Set the center section rigging on all axes.
2) Set the dihedral angle of the front spars of the lower wings.
3) Set the incidence angles of the lower wings by changing the dihedral of the aft wing spars.
4) Set the stagger.
5) Set the dihedral angle of the front spars of the upper wings.
6) Set the incidence angles of the upper wings by changing the dihedral of the aft wing spars.
7) Check the wing-fuselage alignment.
8) Set wash-in and wash-out if applicable.
9) Recheck the rigging at all points.
10) Safety all tie-rods, struts, and adjustments.

Rigging the Center Section

Where the upper wing is composed of three pieces, the center section of the upper wing should be rigged before other sections are attached. Refer to Figure 11-2. The first step is to locate it so that the longitudinal centerline of the center section is directly above the centerline of the ship.

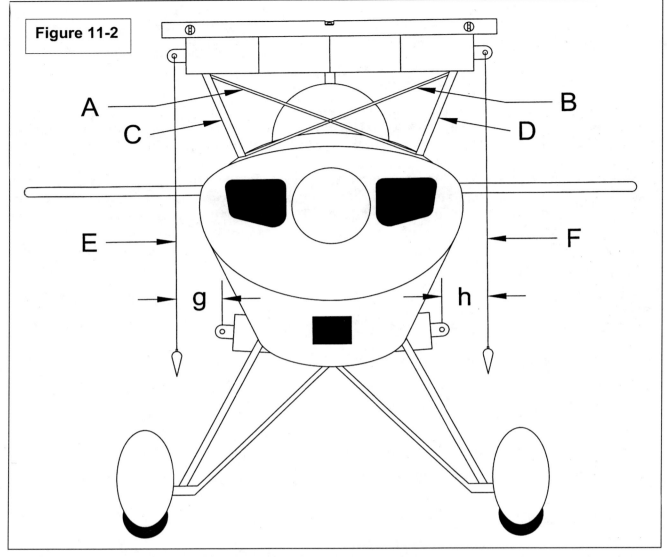

Figure 11-2

If struts (C) and (D) are adjustable for length, make them exactly the same length (a trammel bar is useful here).

Center the wing center section on the fuselage by adjusting the cross wires (A) and (B) until they trammel the same from pin to pin. If the length of the wire (A) is greater than that of wire (B), loosen the wire (B) and tighten the wire (A).

NOTE

When adjusting wires that are in tension, always loosen one wire before tightening the opposing wire.

When wires (A) and (B) are adjusted to equal length the center section is centered properly. To check the accuracy of the work, drop plumb lines (E) and (F) from the upper hinge fittings and measure the horizontal distance from the lines to some structural part that is accurately located, such as the lower hinge fittings. If the fuselage is level laterally and the center section is centered, the distances (g) and (h) will be the same.

An alternative to plumb bobs is to fabricate a rigging tool like in Figure 11-3. The mounting points of the tool on the airplane should be some structural part(s) of the aircraft that have been precisely installed or aligned.

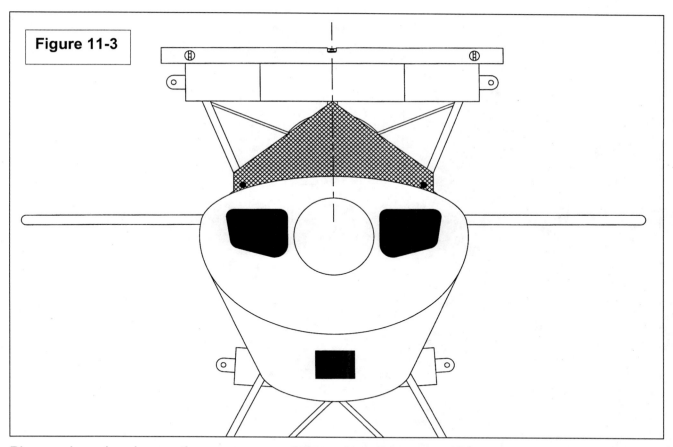

Figure 11-3

Place a long level over the upper spar fittings (using spacers if necessary to raise the level above the rest of the center section, see Chapter 5 on leveling). If the center section is not level, the struts (C) and (D) should be readjusted.

It is possible here that misalignment (center section not level) may be due to improper stagger, illustrated in Figure 11-4. It will be noticed that when the stagger is changed, strut (D) pivots on point (E) and describes an arc, shown by the dotted lines at the top of strut (D). Thus, any change in stagger will affect the horizontal level of the center section. To check this, drop a plumb line (A) over the leading edge of the center section near each end.

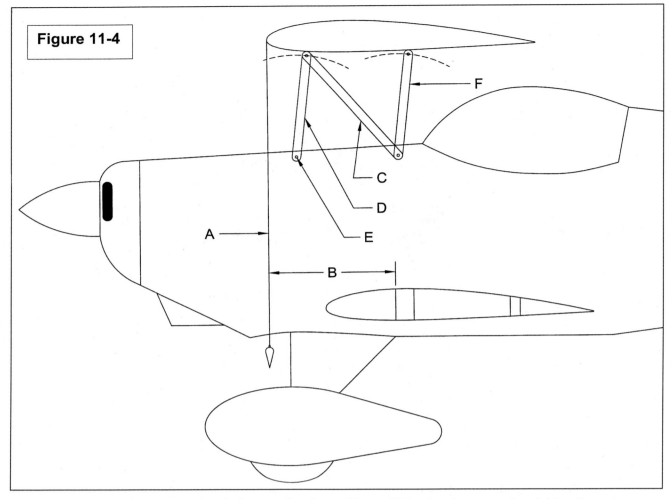

Figure 11-4

The distance (B) from the plumb line to the front hinge fitting/main spar should be the same on both sides. If the lower wing is on, the measurement may be taken from the plumb line to the leading edge of the lower wing. If the stagger is greater on one side than on the other, the stagger struts (C) should be adjusted.

After locating the center section above the centerline of the fuselage, leveling it and adjusting the stagger on both sides, the next step is to set the incidence of the center section. The tools and procedures necessary for making the measurement are the same as for monoplanes, described in Chapter 3. Both ends of the center section should be checked for incidence and any twist removed. The difference should be corrected by lengthening or shortening struts (F).

After the above points have been checked, the rigging of the center section can be regarded as complete. Many center sections do not require as much rigging as has been described, for they may have non-adjustable struts, making it impossible to adjust each angle. In this case the rigging is limited to adjusting the tie-rods. It is suggested to wait until the other wings are installed before safetying all of the connections for the center section, in case any errors are found.

Rigging the Wings

If both wings have dihedral, the lower wings are rigged first. Slack all of the flying wires and set the dihedral. On a conventional biplane the front spar of the wing is usually accepted as being the wing reference line. Dihedral measurements are discussed in other chapters. Refer to Figure 11-5. Tightening the landing wires increases the dihedral and conversely, loosening the landing wires decreases dihedral.

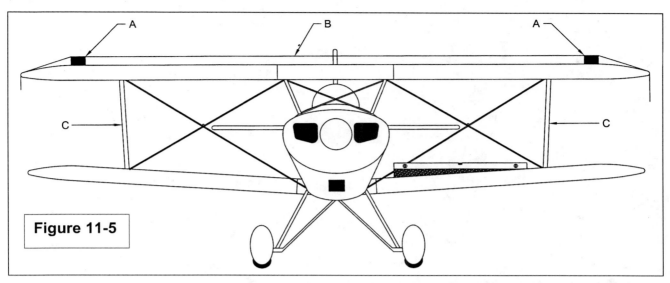

Figure 11-5

When a biplane has no dihedral in the upper wings (as many don't), a simpler procedure is to first rig the top wing straight, or level laterally (this requires that the interplane struts be adjustable for length). The procedure for setting the top wing is similar as in Chapter 5 for a high-wing monoplane. Either the string method or a level may be employed. Where the string method is used and the wing is level (no dihedral), some blocks of equal size are necessary to raise the string above the wing ((A) in Figure 11-5). Adjust the landing wires so that the distances from the string to any corresponding portions of the wings or center section are the same. Then the lower wing dihedral may be adjusted by adjusting the struts (C) in Figure 11-5 so that they are the same length.

The next step in rigging a biplane is to adjust the incidence angle/washout of both lower wings (see Chapter 3 on incidence measurements). If the incidence of the lower wings is not adjustable at the wing root, only the incidence of the wing tips will be adjustable (washout). If the incidence angle at the wing tip is too great, it may be decreased by tightening the rear landing wire. This may also require adjustment in the aft interplane strut.

The stagger of the wings may be checked by dropping a plumb bob over the leading edge of the top wing and measuring the horizontal distance from the line to the leading edge of the lower wing. A convenient method of supporting the plumb line is to tie it to the top of a long pole which can be held above the wing or laid against it, in such a manner that the plumb line rests firmly on the leading edge. If any difficulty is encountered in taking the actual measurement, due to the continuous swinging of the plumb bob, allowing the plumb bob to descend into a pail of water can steady the line. However, care should be taken that it does not touch the bottom or sides of the pail. Inasmuch as the stagger of the center section has been previously adjusted, a mere check should be sufficient at that point. Next, check the stagger at a point directly in front of the interplane strut. If the stagger is too great, decrease the length of the diagonal, or stagger strut (A in Figure 11-6). The stagger can be decreased by shortening the same strut.

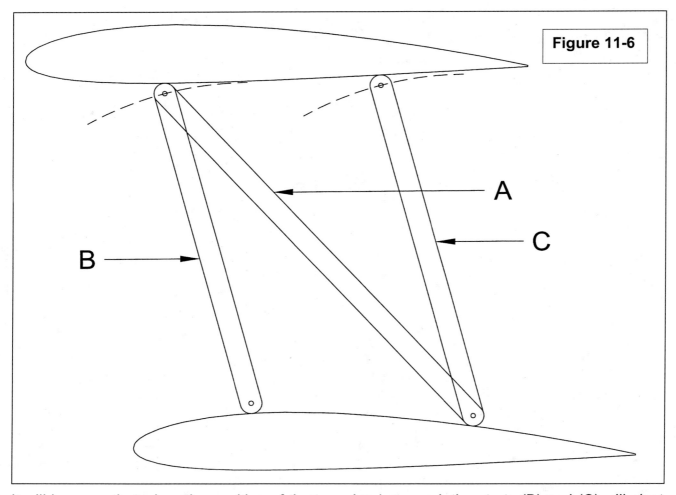

Figure 11-6

It will be seen that when the position of the top wing is moved, the struts (B) and (C) will pivot at their lower ends and their tops will describe arcs indicated by the dotted lines. If the struts and their attachments are so designed that the center lines of the struts are parallel, and the distance between the chord lines of the wings is the same (gap), the incidence angle of the top wing will not be changed by an increase or decrease of stagger. If it is not so designed, an adjustment will be provided in strut (B) or (A), or both. Thus, if an adjustment is provided in strut (B), the next step in the rigging is to adjust the dihedral angle of the front spar of the upper wing. This will be done by increasing or decreasing the length of the front strut (B). Check the incidence angle of the upper wing and adjust it by means of strut (C) so that it is the same at all points. If it is too great at the wing tip it can be decreased by raising the rear spar. In order to raise the rear spar without affecting the previously established stagger, both the stagger strut (A) and the rear strut (C) will have to be lengthened proportional amounts. Conversely, the incidence/washout angle can be increased by shortening both struts. After the above described adjustments have been made, the flying wires can be tightened to the proper tension.

Before proceeding further, the angle which the wings form with the fuselage should be checked to make sure that no undue strain has been placed on any member (Figure 5-7). To do this measure from some corresponding point, such as the interplane strut fittings, to a central point located in the nose of the airplane, like the centerline of the engine mount. The measurement of the right hand side should be the same as that of the left hand side. If a difference of more than 1/4" is discovered, the entire rigging should be re-checked.

The final step in rigging biplane wings is to adjust the wash-in and wash-out. It is common practice in biplanes to put wash-in and wash-out in the bottom wings only. If the exact amount of wash-in/wash-out is not given it is safest not to increase the angle of wing setting more than one degree. It is set by adjusting the length of the rear strut and the rear landing wire. After these final adjustments have been made, all nuts and bolts should be torqued to their final value, strut adjustments should be locked, and all tie-rods streamlined and safetied.

Appendix A
Math for Rigging

Angular units of measurement may appear in several forms;

- degrees-minutes-seconds (47°30'30")
- degrees-decimal minutes (47°30.5')
- decimal degrees (47.5000231°)

One degree is equal to 1/360 of a circle.

One minute (minute of arc, arcminute) is equal to 1/60 of a degree.

One second (second of arc, arcsecond) is equal to 1/60 of a minute.

Present day ISO standards specify that angles should be given in decimal degrees.

Mostly used in mathematics, the radian is equal to 180/pi, or approximately 57.27°.

Finding the Chord of An Arc

Being able to determine the straight line distance of the trailing edge of a control is useful for rigging.

Chord Length, Inches $\quad a = R \times \left(2 \times \sin\left(\dfrac{v}{2}\right) \right)$

where $\quad v = \dfrac{360}{n}$, degrees

and $\quad R = $ radius of the arc, inches

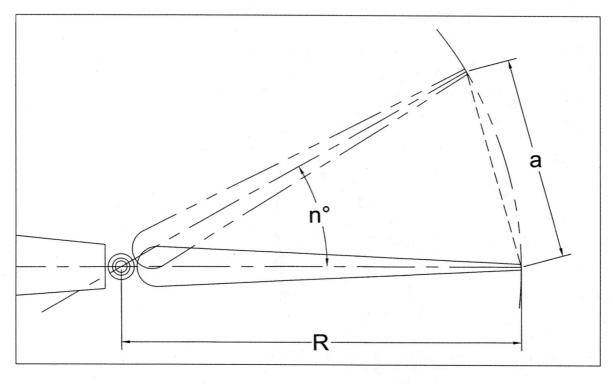

Trigonometric Functions

The trigonometry functions have many uses in layout, construction, and rigging. A cheap scientific calculator will convert trigonometric functions like sine, cosine, etc., into an angle in degrees, and vice-versa (make sure it is set to display degrees rather than radians).

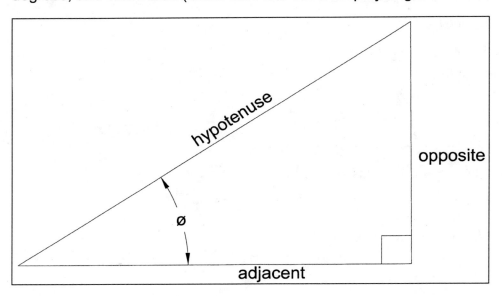

Sine: $$\sin(\theta) = \frac{\text{opposite}}{\text{hypotenuse}}$$

Cosine: $$\cos(\theta) = \frac{\text{adjacent}}{\text{hypotenuse}}$$

Tangent: $$\tan(\theta) = \frac{\text{opposite}}{\text{adjacent}}$$

References

1. Design of Light Aircraft, R. D. Hiscocks
2. ACES Guide to Propeller Balancing, TEC Aviation Division
3. Aircraft Weight and Balance Handbook (FAA-H-8083-1), Federal Aviation Administration
4. Custom Built Sport Aircraft Handbook, Experimental Aircraft Association
5. Techniques of Aircraft Building, Experimental Aircraft Association
6. AC 65-9A, A & P Mechanics General Handbook, Federal Aviation Administration
7. AC 65-12A, A & P Mechanics Powerplant Handbook, Federal Aviation Administration
8. AC 65-15A, A & P Mechanics Airframe Handbook, Federal Aviation Administration
9. AC 43.13-1B, Acceptable Methods, Techniques and Practices - Aircraft Inspection and Repair, Federal Aviation Administration
10. AC 43.13-2A, Acceptable Methods, Techniques and Practices - Aircraft Alterations, Federal Aviation Administration
11. AC 20-37E, Metal Propeller Maintenance, Federal Aviation Administration
12. NACA/NASA Report No. XXX, National Advisory Committee for Aeronautics and National Aeronautics and Space Administration (NOTE: Numerous NASA and NACA technical reports and research papers were used in the compilation of this book. They are available free for free download at the NASA Technical Reports Server. Since the 1920's, they have produced thousands of research papers on aerodynamics, structures, flying qualities and techniques, and specific aircraft configuration problems. Most textbooks are very general in nature and do not address specific aircraft configurations and controls. A careful search of the available research material will probably produce a large quantity of exactly what one is looking for, for a specific problem. NACA was the forerunner of NASA and much of the most useful material they produced was in the 1940's and 1950's, the heyday of propeller driven aircraft. It takes some practice to quickly search through the research materials as the terminology has changed over the years.)
13. ANC-12, Vibration and Flutter Prevention Handbook, Air Force-Navy-Civil Committee on Aircraft Design Criteria
14. Numerous aircraft manufacturers maintenance manuals and Type Certificate Data Sheets
15. Airplane Design, Volumes 1-7, Jan Roskam
16. Code of Federal Regulations, Title 14-Chapter 1-Subchapter C-Part 23, Federal Aviation Administration
17. Incidence is Not Incidental, Noel J. Becar, Sport Aviation, March 1964
18. Airframe and Equipment Engineering Report No. 45, Simplified Flutter Prevention Criteria for Personal Type Aircraft, Federal Aviation Administration
19. BeDesign Notes No.1, Bede Aircraft Inc.
20. Landing Gear Design For Light Aircraft, Volume 1, Ladislao Pazmany
21. Aircraft Design: A Conceptual Approach, Daniel P. Raymer
22. Pilot's Handbook of Aeronautical Knowledge (FAA-H-8083-25), Federal Aviation Administration

23. Wood Propellers: Installation, Operation, and Maintenance, Sensenich Wood Propeller Company
24. Aerodynamics for Naval Aviators NAVWEPS 00-80T-80, H.H. Hurt, Jr.
25. Airplane Flight Dynamics and Automatic Flight Controls, Part 1, Jan Roskam
26. Mechanical Vibrations, W.T. Thomson

also from David Russo

Construction of Tubular Steel Fuselages

by David Russo; Vex Aviation
ISBN 0-9774896-0-4

This book is written to assist the average mechanic or aircraft builder to produce a tubular steel airframe that is as good as one from the factory. The assembly methods in this manual are intended to produce a highly accurate fuselage in such a way that expansion and contraction occur symmetrically with no distortion. The techniques given including those on jigging and alignment methods will also help in rapid prototyping where a high degree of accuracy is necessary without elaborate welding jigs.

This book details the entire fuselage design and construction process in a practical way for the builder, restorer, or mechanic (not the engineer) including consideration for loads, evaluation and placement of fittings and joints, jigging, welding and other metal working techniques, tool use, and the actual step by step construction of top and bottom halves of a steel fuselage as well as the many sub assemblies common to every tubular aircraft fuselage type from J3 replicas to high performance aerobatic aircraft.

book or e-Book $34.95

order from:

PO Box 270   Tabernash CO 80478
800 780-4115   fax 970 887-2197
www.ACtechbooks.com

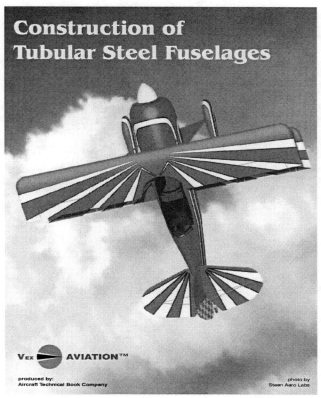

TABLE OF CONTENTS
Welded Fuselage Construction and Design
terminology, axes, fuselage geometry, drawing conventions, materials

Methods & Processes
abrasives, welding nonstructural attach points, corrosion removal, painting

Jig Tables
flat and level, materials, order of construction

Layout
centerlines, drawing tops and bottoms

Constructing the Top and Bottom
blocks, shims, longerons, constructing the top, constructing the bottom

Fitting the Top and Bottom
hangers, reference plates, crossmembers, longerons

Welding the Fuselage
welding the joints, longeron straightening caused by welding, fabricating engine mount pads, finishing welding, rudder post

Horizontal/Vertical Stabilizer and Control Surfaces

Seats, Rollover Structures, Harness Installations

notes